The Human Brain during the Third Trimester 225- to 235-mm Crown-Rump Lengths

This eleventh of 15 short atlases reimagines the classic 5-volume *Atlas of Human Central Nervous System Development*. This volume presents serial sections from specimens between 225 mm and 235 mm with detailed annotations. An introduction summarizes human CNS developmental highlights around 6.5 months of gestation. The Glossary (available separately) gives definitions for all the terms used in this volume and all the others in the *Atlas*.

Key Features

- Classic anatomical atlases
- Detailed labeling of structures in the developing brain offers updated terminology and the identification of unique developmental features, such as germinal matrices of specific neuronal populations and migratory streams of young neurons
- Appeals to neuroanatomists, developmental biologists, and clinical practitioners
- A valuable reference work on brain development that will be relevant for decades

ATLAS OF
HUMAN CENTRAL NERVOUS SYSTEM DEVELOPMENT
Series

Volume 1: **The Human Brain during the First Trimester 3.5- to 4.5-mm Crown-Rump Lengths**

Volume 2: **The Human Brain during the First Trimester 6.3- to 10.5-mm Crown-Rump Lengths**

Volume 3: **The Human Brain during the First Trimester 15- to 18-mm Crown-Rump Lengths**

Volume 4: **The Human Brain during the First Trimester 21- to 23-mm Crown-Rump Lengths**

Volume 5: **The Human Brain during the First Trimester 31- to 33-mm Crown-Rump Lengths**

Volume 6: **The Human Brain during the First Trimester 40- to 42-mm Crown-Rump Lengths**

Volume 7: **The Human Brain during the First Trimester 57- to 60-mm Crown-Rump Lengths**

Volume 8: **The Human Brain during the Second Trimester 96- to 150-mm Crown-Rump Lengths**

Volume 9: **The Human Brain during the Second Trimester 160- to 170-mm Crown-Rump Lengths**

Volume 10: **The Human Brain during the Second Trimester 190- to 210-mm Crown-Rump Lengths**

Volume 11: **The Human Brain during the Third Trimester 225- to 235-mm Crown-Rump Lengths**

Volume 12: **The Human Brain during the Third Trimester 260- to 270-mm Crown-Rump Lengths**

Volume 13: **The Human Brain during the Third Trimester 310- to 350-mm Crown-Rump Lengths**

Volume 14: **The Spinal Cord during the First Trimester**

Volume 15: **The Spinal Cord during the Second and Third Trimesters and the Early Postnatal Period**

The Human Brain during the Third Trimester 225- to 235-mm Crown-Rump Lengths

Atlas of Human Central Nervous System Development, Volume 11

Shirley A. Bayer
Joseph Altman

CRC Press
Taylor & Francis Group
Boca Raton London New York

CRC Press is an imprint of the
Taylor & Francis Group, an **informa** business

Designed cover: Shirley A. Bayer and Joseph Altman

First edition published 2024
by CRC Press
2385 NW Executive Center Drive, Suite 320, Boca Raton, FL 33431

and by CRC Press
4 Park Square, Milton Park, Abingdon, Oxon, OX14 4RN

CRC Press is an imprint of Taylor & Francis Group, LLC

LCCN no. 2022008216

ISBN: 978-1-032-22874-7 (hbk)
ISBN: 978-1-032-22004-8 (pbk)
ISBN: 978-1-003-27460-5 (ebk)

DOI: 10.1201/9781003274605

Typeset in Times Roman
by KnowledgeWorks Global Ltd.

Publisher's note: This book has been prepared from camera-ready copy provided by the authors.
Access the Support Material: www.routledge.com/9781032228747

CONTENTS

ACKNOWLEDGMENTS

We thank the late Dr. William DeMyer, pediatric neurologist at Indiana University Medical Center, for access to his personal library on human CNS development. We also thank the staff of the National Museum of Health and Medicine that were at the Armed Forces Institute of Pathology, Walter Reed Hospital, Washington, D.C. when we collected data in 1995 and 1996: Dr. Adrianne Noe, Director; Archibald J. Fobbs, Curator of the Yakovlev Collection; Elizabeth C. Lockett; and William Discher. We are most grateful to the late Dr. James M. Petras at the Walter Reed Institute of Research who made his darkroom facilities available so that we could develop all the photomicrographs on location rather than in our laboratory in Indiana. Finally, we thank Chuck Crumly, Neha Bhatt, Kara Roberts, Michele Dimont, and Rebecca Condit for expert help during production of the manuscript.

AUTHORS

Shirley A. Bayer received her PhD from Purdue University in 1974 and spent most of her scientific career working with Joseph Altman. She was a professor of biology at Indiana-Purdue University in Indianapolis for several years, where she taught courses in human anatomy and developmental neurobiology while continuing to do research in brain development. Her lengthy publication record of dozens of peer-reviewed, scientific journal articles extends back to the mid 1970s. She has co-authored several books and many articles with her late spouse, Joseph Altman. It was her research (published in *Science* in 1982) that proved that new neurons are added to granule cells in the dentate gyrus during adult life, a unique neuronal population that grows. That paper stimulated interest in the dormant field of adult neurogenesis.

Joseph Altman, now deceased, was born in Hungary and migrated with his family via Germany and Australia to the US. In New York, he became a graduate student in psychology in the laboratory of Hans-Lukas Teuber, earning a PhD in 1959 from New York University. He was a postdoctoral fellow at Columbia University, and later joined the faculty at the Massachusetts Institute of Technology. In 1968, he accepted a position as a professor of biology at Purdue University. During his career, he collaborated closely with Shirley A. Bayer. From the early 1960s-2016, he published many articles in peer-reviewed journals, books, monographs, and free online books that emphasized developmental processes in brain anatomy and function. His most important discovery was adult neurogenesis, the creation of new neurons in the adult brain. This discovery was made in the early 1960s while he was based at MIT, but was largely ignored in favor of the prevailing dogma that neurogenesis is limited to prenatal development. After Dr. Bayer's paper proved new neurons are added to granule cells in the hippocampus, Dr. Altman's monumental discovery became more accepted. During the 1990s, new researchers "rediscovered" and confirmed his original finding. Adult neurogenesis has recently been proven to occur in the dentate gyrus, olfactory bulb, and striatum through the measurement of Carbon-14—the levels of which changed during nuclear bomb testing throughout the 20th century—in postmortem human brains. Today, many laboratories around the world are continuing to study the importance of adult neurogenesis in brain function. In 2011, Dr. Altman was awarded the Prince of Asturias Award, an annual prize given in Spain by the Prince of Asturias Foundation to individuals, entities, or organizations globally who make notable achievements in the sciences, humanities, and public affairs. In 2012, he received the International Prize for Biology - an annual award from the Japan Society for the Promotion of Science (JSPS) for "outstanding contribution to the advancement of research in fundamental biology." This Prize is one of the most prestigious honors a scientist can receive. When Dr. Altman died in 2016, Dr. Bayer continued the work they started over 50 years ago. In her late husband's honor, she created the Altman Prize, awarded each year by JSPS to an outstanding young researcher in developmental neuroscience.

INTRODUCTION

A. Specimens and Organization

Volume 11 in this Atlas Series presents the human brain in two normal specimens from the Yakovlev Collection[1] at 6.5 months during the early third trimester. These specimens were analyzed in Volume 2 of the original *Atlas of Human Central Nervous System Development* (Bayer and Altman, 2004). By this time, most fetuses are viable *ex utero*. Nearly all the structures present in the adult brain are recognizable and are maturing from the diencephalon to the medulla. But remnants of the embryonic nervous system remain, mainly in the cerebral cortex and in the cerebellar cortex.

This volume contains serial grayscale photographs of Nissl-stained sections of a sagittal specimen (Y147-63, **Part II**) and a horizontal specimen (Y16-59, **Part III**) with crown-rump (CR) lengths of 225 mm and 235 mm, respectively. Both specimens are approximately in the 26th gestational week (GW).

Sagittal Plates are ordered from medial to lateral; the anterior part of each photographed section is facing left, posterior right. **Horizontal Plates** are presented from dorsal (first) to ventral (last); the anterior part of each section faces left, the posterior part faces right; the midline is in the vertical center. Each **Plate** is in two parts: **A**, on the left, shows the full-contrast photograph without labels; **B,** on the right, shows a low-contrast copy of each photograph with unabbreviated labels. For each specimen, a series of serially spaced **low-magnification plates** show the entire section to identify large structures. The brain core is shown in many **high-magnification plates** to identify smaller structures. In addition, several **very-high -magnification plates** show the cerebral and cerebellar cortices. Because our emphasis is on development, transient structures that appear only in immature brains are labeled in ***italics***, either directly in some of the high-magnification plates or in **bold numbers** that refer to labels in a list. During dissection, embedding, cutting, and staining, some of the sections illustrated were torn; that damage is sometimes surrounded by *dashed lines*.

1. The *Yakovlev Collection* (designated by a **Y** prefix in the specimen number) is the work of Dr. Paul Ivan Yakovlev (1894–1983), a neurologist affiliated with Harvard University. Throughout his career, Yakovlev collected many diseased and normal human brains. He invented a giant microtome that was capable of sectioning entire human brains. Later, he became interested in the developing brain and collected many during the second and third trimesters. The normal brains in the developmental group were cataloged by Haleem (1990) and were examined by us during 1996 and 1997. The collection was moved to the National Museum of Health and Medicine when the Armed Forces Institute of Pathology (AFIP) closed at Walter Reed Hospital and is still available for research.

B. Developmental Highlights

Figures 1-2 compare second trimester and early third trimester brains in horizontal sections of the dorsal telencephalon. The overall growth of the brain is obvious when the sections are stacked in a "layer cake" arrangement (**Fig. 1**). The side-by-side comparison (**Fig. 2**) shows that parts of the cerebral cortex are growing at different rates. From GW 20 to GW 23 to GW 26, the cortical plate gradually increases its thickness and rapidly grows lengthwise. The white matter (*pale yellow*) greatly increases its thickness in most cortical areas except the insula. On the other hand, layers 3-6 of the *stratified transitional field* slow their growth, especially between GW 20 and GW 23; by GW 26, these layers have nearly disappeared in the frontal areas and are very thin in posterior (occipital) areas. Synchronously with these growth dynamics, the cortex begins to fold outward (gyri) and inward (sulci and fissures) to accommodate the rapidly increasing surface area.

Cortical gyrification is an intriguing phenomenon that neuroanatomists have studied for many years; Welker (1990) wrote a comprehensive review of the literature in mammals (including humans). The abstract of his review is that uneven expansion of cortical areas cause gyri and sulci to develop, with

Superimposed Early Third Trimester and Second Trimester Brains in the Horizontal Plane

Bottom layer: GW 26
Middle layer: GW 23
Top layer: GW 20

NEP- neuroepithelium
STF-stratified transitional field
SVZ-subventricular zone

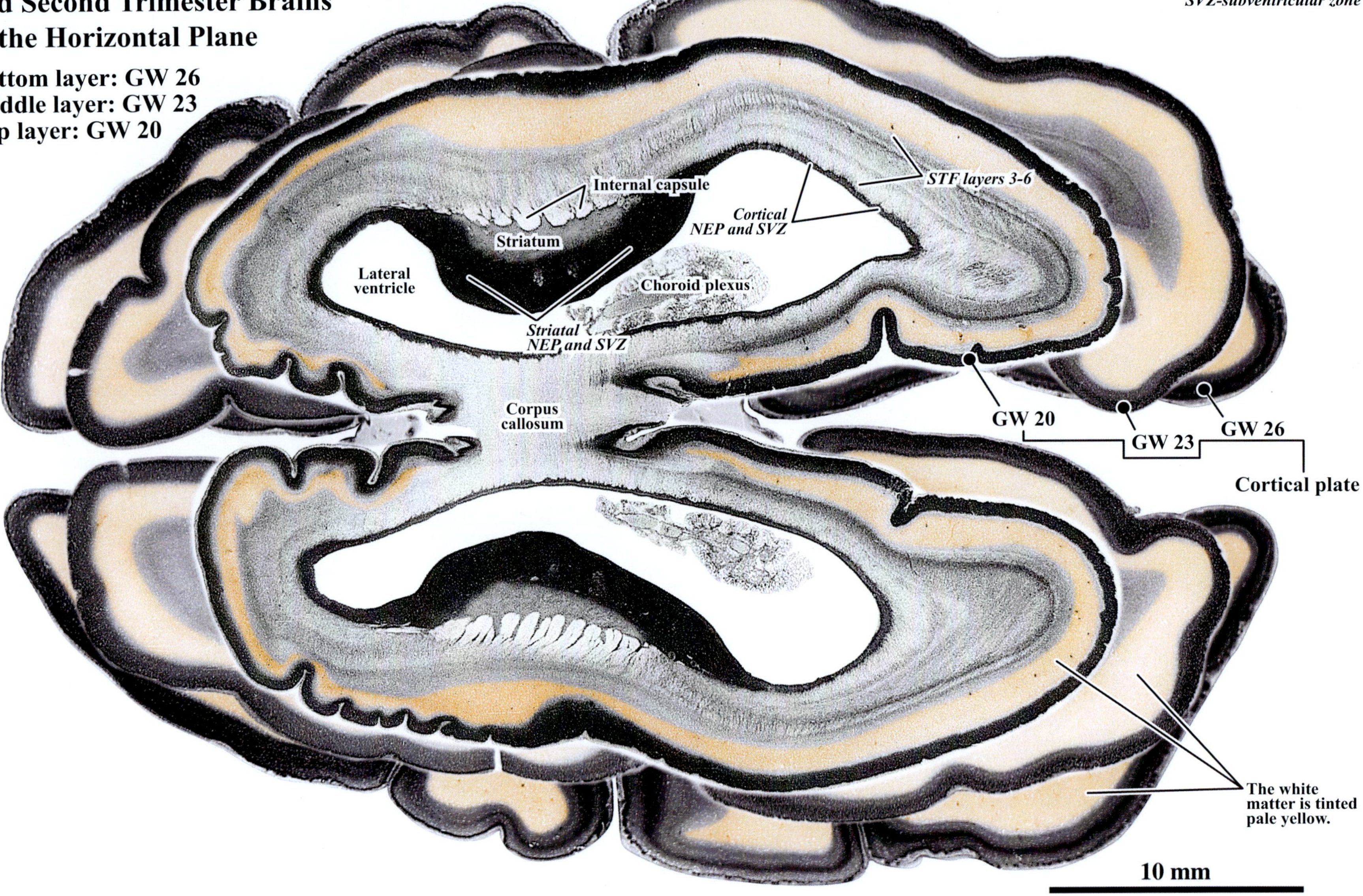

Figures 1 and 2 (*facing pages*). A comparison of second and early third trimester brains in the horizontal plane to show the relative growth of different components of the cerebral cortex. Note that the internal capsule and corpus callosum are well established by GW 20. Massive growth of fibers into the white matter (*pale yellow*) from subcortical areas (mainly the thalamus) and from neurons within the cortical plate growing axons into the white matter (some extend to the opposite side in the corpus callosum) are related to the formation of fissures, sulci, and gyri. (GW 20 is shown in Plate 41, Volume 9, Bayer and Altman, in Press; GW 23 in Plate 1, Volume 10, Bayer and Altman, in Press; GW 26 in Plate 28, this Volume).

Side-by-Side Comparison of Early Third Trimester and Second Trimester Brains in the Horizontal Plane

Growth related to increased gyrification:

Cortical plate increases thickness and linear surface

White matter increases depth in most areas (insula excepted)

STF layers 3-6 decrease in relative size

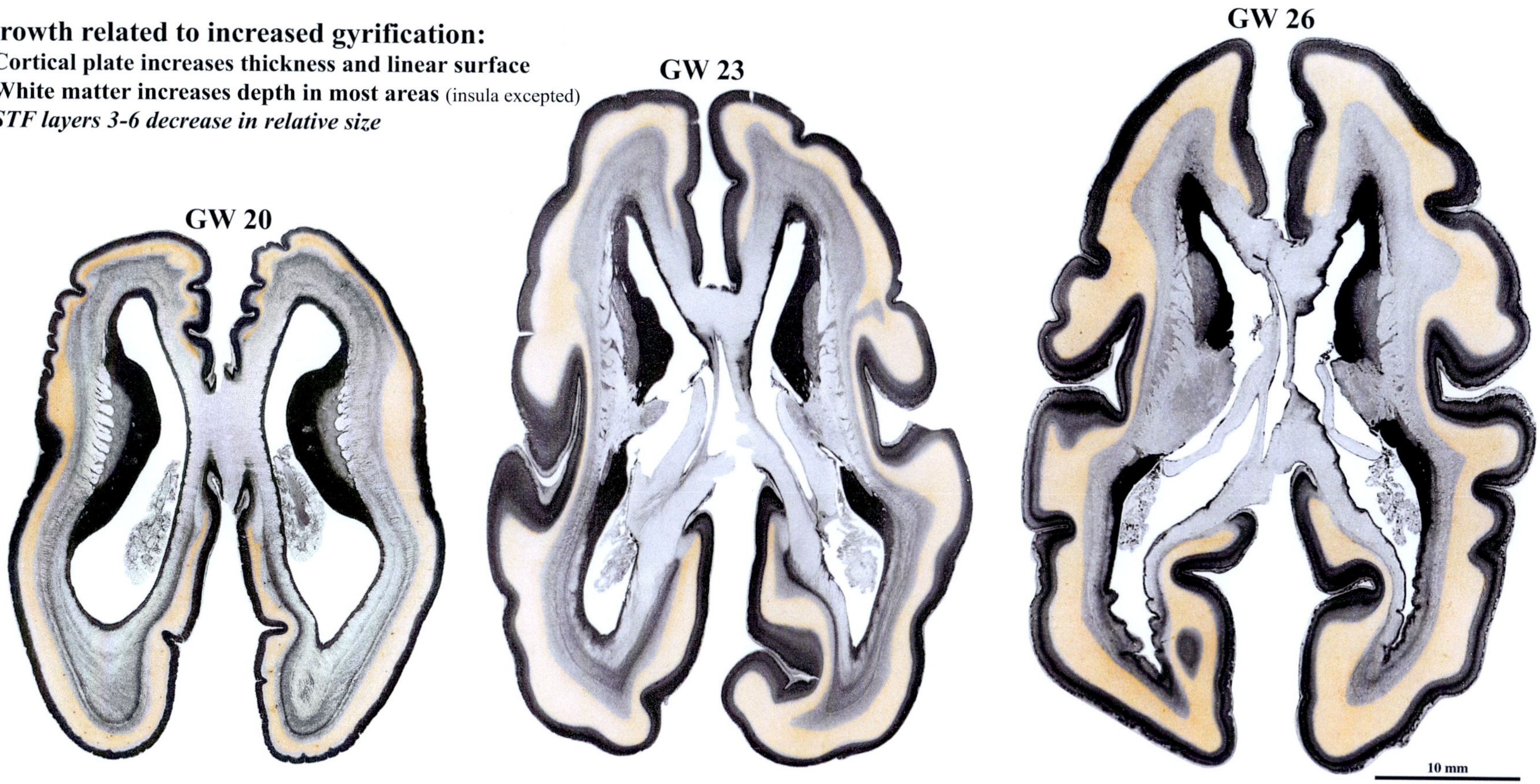

slower- growing parts of cortex in the fundus of a sulcus and faster-growing cortex in the walls and the crowns of a gyrus. Our observations concur that uneven growth results in gyrification and also puts forth the hypothesis that the pattern of cortical gyrification results from the establishment of early anatomical connections with specific areas of the cortical plate (Altman and Bayer, 2015). Our hypothesis is that early connections with the thalamus "anchor" specific areas of the cortical plate (*see* Fig. 2, Volume 10, Bayer and Altman, in press, also Van Essen, 1997) and initiates the process of gyrification. Thalamic neurogenesis is completed during the middle part of the first trimester, and it is plausible that some early contacts are made that set off unique cortical areas. Furthermore, thalamic axons are a main component of the internal capsule fiber tract[2] that first appears on GW 8.4 (*see* Plates 20-21, Volume 4, Bayer and Altman, 2023)

2. The internal capsule is a major tract in the telencephalon that links the cortex to subcortical structures. Some of its components are afferent axons from the brainstem (e.g., locus coeruleus), thalamo-cortical axons, cortico-thalamic axons, axons that will form the corticospinal and corticobulbar tracts, and axons that run to and from the basal ganglia.

The first anchoring occurs with the ventrolateral cortical plate that will become the insula at the base of the lateral fissure. The insula divides the lateral cortex into an anterior frontal lobe and a posterior temporal lobe. The cortical plate appears first in the future insula on GW 8.4 (*see* Plates 19-20, Volume 4, Bayer and Altman, 2022). By GW 9.6, the cortical plate is extensive throughout the entire cerebral wall but is most thick in a central lateral area (insula) that has a slight indentation. This part of the cortex is directly adjacent to the entry zone of the massive internal capsule (*see* Plate 5, Volume 5, Bayer and Altman, 2023).

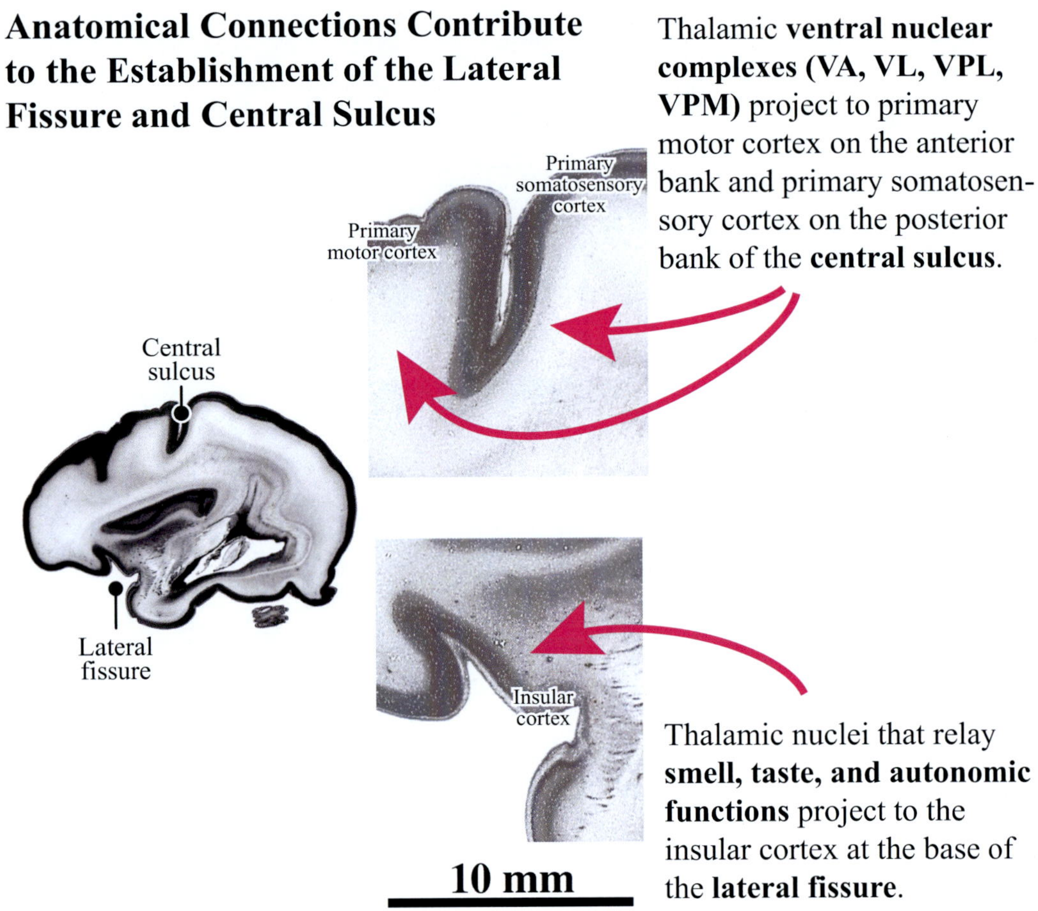

Figure 3. A lateral sagittal section of the brain of Y147-63 from **Plate 9** in **Part II** of this volume is used to indicate the anatomical connections of the sensorimotor cortex (*top*) around the central sulcus and some anatomical connections of the insula (*bottom*). Abbreviations: VA, ventral anterior nucleus (relays input from the globus pallidus); VL, ventral lateral nucleus (relays input from the cerebellum); VPL, ventral posterolateral nucleus (relays touch and pain from the medial lemniscus and the spinothalamic tracts from the lower limbs and trunk); VPM, ventral posteromedial nucleus (relays touch and pain from the medial lemniscus and the spinothalamic tracts from the upper limbs and trunk).

All parts of the mature insular cortex are contacted by thalamic nuclei, especially those related to olfaction, taste, and autonomic functions (**Fig. 3**, Ghaziri et al., 2018).

The next cortical area to be cordoned off is the primary visual cortex in the depths of the calcarine fissure (**Fig. 4**). On GW 14, a sharp and deep indent occurs in the posteriomedial cerebral wall (*see* Plates 10-13, Volume 8, Bayer and Altman, in press), the presumptive calcarine sulcus. Already by the end of the first trimester, some internal capsule fibers can be seen curving sharply back toward the posterior cortex (*see* Plates 5-6, Volume 7, Bayer and Altman, 2023). The lateral geniculate body in the thalamus gets point-to-point input from retinal ganglion cells and then relays that ordered input to both sides of the calcarine sulcus so that the lower visual field maps to the cuneus and the upper visual field maps to the lingula.

The parieto-occipital sulcus appears slightly later than the calcarine sulcus, and the two form an L-shaped cleft in the medial cortical wall in sagittal sections (**Fig. 4**). The anterior side of the sulcus (visual area 6) gets strong input from the large pulvinar nucleus in the thalamus that relays secondary visual information as well as other sensory modalities (Gamberini, et al. 2015). It is our hypothesis that initial connections with this part of the cortex contributes to the establishment of the sulcus and the posterior extent of the parietal lobe.

The paracentral lobule that surrounds both sides of the central sulcus (**Fig. 3**) appears first on GW 20 in the ventrolateral cortical plate (*see* Plates 15 and 36, Volume 9, Bayer and Altman, in press). That corresponds to our findings in rats (Bayer and Altman, 1991), now also in humans (the current Atlas Series), that ventrolateral cortical areas develop earlier than dorsomedial areas. The central sulcus is prominent in lateral sagittal sections (**Fig. 3**), and absent in medial sections (*compare* **Plates 4 and 10 in Part II** of this Volume). It is well known that the large ventral nuclear complexes in the thalamus project to the primary motor area on the anterior bank of the central sulcus (for a review and a recent current study, *see* Rouiller, et al., 1999), and the same thalamic areas also project to the primary somatosensory cortex on the posterior bank of the sulcus (Huffman and Krubitzer, 2001). These projections are topographically organized and produce a map of the body parts (the homunculus) on both primary motor and primary somatosensory cortices. There is a homunculus in rats as well, and we found that the pattern of anatomical connections with the thalamus can be linked to the neurogenetic gradients in both the thalamus and the cortex (Bayer and Altman, 1991).

By the beginning of the third trimester, all major gyri and sulci have appeared and the vast cortical sheet is presumably shaped by early connections with the thalamus. The result is seven anatomical areas.

First, the limbic lobe ("edge" lobe) was not discussed, but much of it is established by GW 7.4 as the hippocampus becomes definite in the dorsomedial edge of the cerebral cortex adjacent to the presumptive cingulate cortex (see Plates 24-25, Volume 3, Bayer and Altman 2023). That is before the cortical plate appears and is due to intrinsic gene expression of the LIM-homeodomain gene *Lhx2* (Bulchand et al., 2001). Growth of the entire cortex causes the ventral hippocampus and the parahippocampal cortex to be displaced near the insula (*see* Plates 10 and 11, Volume 6, Bayer and Altman, 2023). **Second and third**, the lateral frontal and the temporal lobes are established by the end of the first trimester. They grow backward and forward respectively to cover the insula during the second trimester. **Fourth**, the primary visual cortex and secondary visual cortex in the occipital lobe around the calcarine and parieto-occipital sulci appear in the early second trimester. **Fifth**, the paracentral lobule surrounding the central sulcus also appears in the second trimester ventrolaterally and gradually extends dorsomedially during the third trimester. **Sixth**, the medial frontal lobe anterior to the precentral gyrus is cordoned off in the late second trimester. **Seventh**, the medial parietal lobe posterior to the postcentral gyrus is cordoned off in the late second trimester. Secondary and tertiary gyrification will continue through the third trimester, after birth, into infancy, childhood, juvenile ages, and not come to an end until the young adult period.

Anatomical Connections Contribute to the Establishment of the Calcarine and Parieto-occipital Sulci

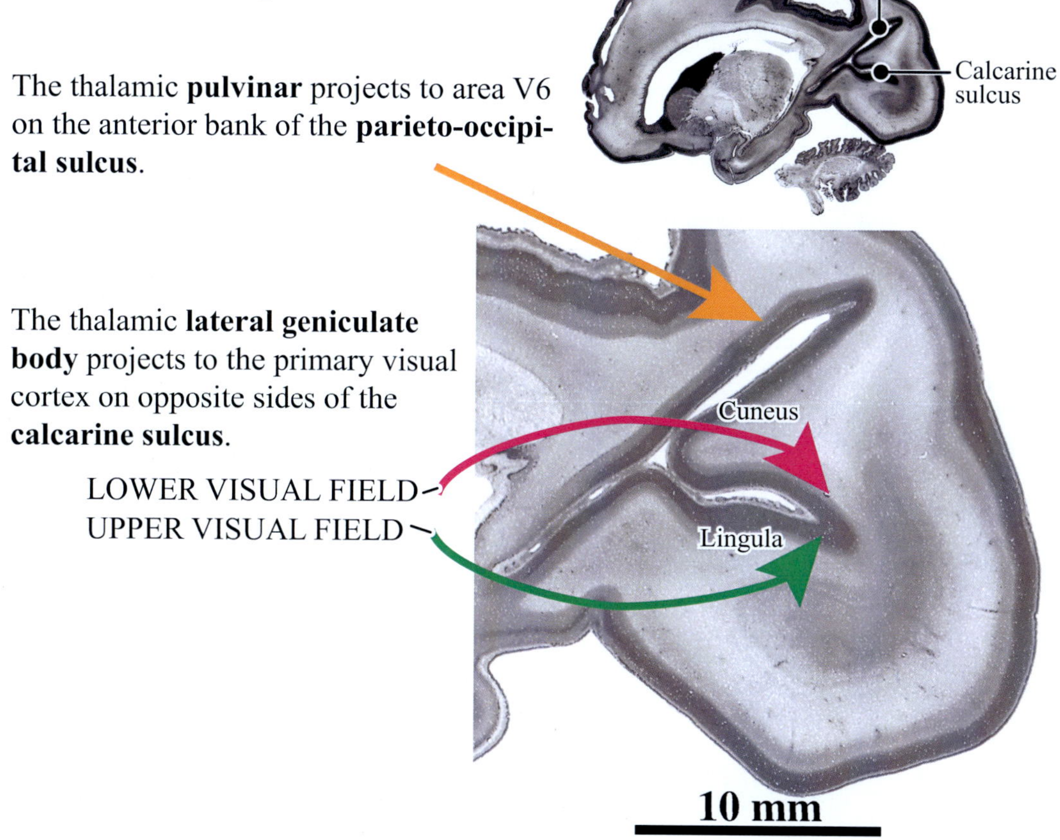

Figure 4. A midlateral sagittal section of the brain of Y147-63 from **Plate 5 in Part II** of this volume is used to indicate the anatomical connections surrounding the parieto-occipital sulcus (*top orange arrow*) and the point-to-point input of the primary visual cortex surrounding the calcarine sulcus (*bottom pink and green arrows*). The upper visual field in the lingula maps the individual's world-centered space, while the lower visual field in the cuneus maps the individual's egocentric space. Visual association cortex around the parieto-occipital sulcus contains areas involved with eye-hand coordination during reaching and grasping (*see* Figure 77 in Altman and Bayer, 2015). Note that the central sulcus is less deep here than in the section in **Figure 3**.

REFERENCES

Altman J, Bayer SA. (2015) *Development of the Human Neocortex.* Ocala, FL, Laboratory of Developmental Neurobiology, neurondevelopment.org.

Bayer SA, Altman J (1991) *Neocortical Development*. New York: Raven Press.

Bayer SA, Altman J (2004) *Atlas of Human Central Nervous System Development*, Volume 2: *The Human Bran during the Third Trimester.* Boca Raton, FL, CRC Press.

Bayer SA, Altman J (2023) *The Human Brain during the First Trimester 15- to 18-mm Crown-Rump Lengths, Atlas of Human Central Nervous System Development*, Volume 3. Taylor and Francis, CRC Press.

Bayer SA, Altman J (2023) *The Human Brain during the First Trimester 21- to 23-mm Crown-Rump Lengths, Atlas of Human Central Nervous System Development*, Volume 4. Taylor and Francis, CRC Press.

Bayer SA, Altman J (2023) *The Human Brain during the First Trimester 31- to 33-mm Crown-Rump Lengths, Atlas of Human Central Nervous System Development*, Volume 5. Taylor and Francis, CRC Press.

Bayer SA, Altman J (2023) *The Human Brain during the First Trimester 40- to 42-mm Crown-Rump Lengths, Atlas of Human Central Nervous System Development*, Volume 6. Taylor and Francis, CRC Press.

Bayer SA, Altman J (2023) *The Human Brain during the First Trimester 57- to 60-mm Crown-Rump Lengths, Atlas of Human Central Nervous System Development*, Volume 7. Taylor and Francis, CRC Press.

Bayer SA, Altman J (2023) *The Human Brain during the Second Trimester 96- to 150-mm Crown-Rump Lengths, Atlas of Human Central Nervous System Development*, Volume 8. Taylor and Francis, CRC Press.

Bayer SA, Altman J (in press) *The Human Brain during the Second Trimester 160- to 170-mm Crown-Rump Lengths, Atlas of Human Central Nervous System Development*, Volume 9. Taylor and Francis, CRC Press.

Bayer SA, Altman J (in press) *The Human Brain during the Second Trimester 190- to 210-mm Crown-Rump Lengths, Atlas of Human Central Nervous System Development*, Volume 10. Taylor and Francis, CRC Press.

Bulchand S, Grove EA, Porter JD Tole S. (2001) LIM-homeodomain gene *Lhx2* regulates the formation of the cortical hem. *Mechanisms of Development*, 100:165-175.

Curtis BA, Jacobson S, Marcus EM (1972) *An Introduction to the Neurosciences*, Philadelphia: W. B. Saunders.

Gamberini M, Bakola S, Passarelli L, et al. (2015) Thalamic projections to visual and visuomotor areas (V6 and V6a in the rostral bank of the parieto-occipital sulcus of the macaque. *Brain Structure and Function* 221:1573-1589

Ghaziri J, Tucholka A, Girard G, et al. (2018) Subcortical structural connectivity of insular subregions. *Scientific Reports* 8:8596, www.nature.com/scientificreports

Haleem M (1990) *Diagnostic Categories of the Yakovlev Collection of Normal and Pathological Anatomy and Development of the Brain.* Washington, D.C. Armed Forces Institute of Pathology.

Huffman KJ, Krubitzer L (2001) Thalamo-cortical connections of areas 3a and M1 in marmoset monkeys. *Journal of Comparative Neurology* 435:291-310.

Larroche JC (1966) The development of the central nervous system during intrauterine life. In: *Human Development*, F. Falkner (ed.), Philadelphia: W. B. Saunders, pages 257-276.

Rouiller EM, Tanne J, Moret V, Boussaoud D (1999) Origin of thalamic inputs to the primary, premotor, and supplementary motor cortical areas and to area 46 in macaque monkeys: a multiple retrograde tracing study. *Journal of Comparative Neurology,* 409:131-152.

Van Essen DC (1997) A tension-based theory of morphogenesis and compact wiring in the central nervous system. *Nature*, 385:313-318.

Welker W. (1990) Why does the cerebral cortex fissure and fold? A review of determinants of gyri and sulci. In: *Cerebral Cortex*, Jones E G et al., eds. New York: Springer Science and Business Media, pages 3-136.

PART II: Y147-63
CR 225 mm (GW 26)
Sagittal

This specimen is case number W-147-63 (Perinatal RPSL) in the Yakovlev Collection. A female infant survived for 27 days after a premature birth. Death occurred from a hemorrhage in the abdominal cavity. The brain was cut in the sagittal plane in 35-μm thick sections and is classified as a Normative Control in the Yakovlev Collection (Haleem, 1990). Since there is no photograph of this brain before it was embedded and cut, the photograph of the medial view of a GW 25 brain that Larroche published in 1966 (**Figure 5**) is used to show gross anatomical features.

Photographs of 10 different Nissl-stained sections are shown in full in **Plates 1-10**. High-magnification views of the brain core and cerebellum are featured in **Plates 11-20**. Very-high-magnification views of different regions of the cerebellar cortex are shown in **Plates 21-26**. Because the section numbers decrease from **Plate 1** (most medial) to **Plate 10** (most lateral), they are from the left side of the brain; the right side has higher section numbers proceeding medial to lateral. The cutting plane of this brain is nearly parallel to the midline in anterior and posterior parts of each section, including different parts of the cortex. The sections chosen for illustration are spaced closer together near the midline to show small structures in the diencephalon, midbrain, pons, and medulla.

Y147-63 contains several immature structures. In the telencephalon, remnants of the germinal zones are present in all lobes of the cerebral cortex where the ***neuroepithelium/subventricular zone*** is generating neocortical neurons. Migrating and sojourning neurons and/or glia are visible in all lobes of the cerebral cortex in ***stratified transitional fields***, thin but with prominent layering in the occipital lobe, and thicker with faint layering in the frontal, parietal, and temporal lobes. More neurons, glia, and their mitotic precursor cells are migrating through the olfactory peduncle toward the olfactory bulb (***rostral migratory stream***) from a presumed source area in the germinal matrix at the junction between the cerebral cortex, striatum, and nucleus accumbens. Streams of neurons and glia percolate through the claustrum, endopiriform nucleus, external capsule, and uncinate fasciculus in the ***lateral migratory stream***. These cells appear to be heading toward the insular cortex, primary olfactory cortex, temporal cortex, and basolateral parts of the amygdaloid complex. In the basal ganglia, there is a thick ***neuroepithelium/subventricular zone*** overlying the striatum and nucleus accumbens where neurons and glia are being generated; some of these, especially from the accumbal area, will enter the ***rostral migratory stream***. Another region of active neurogenesis in the telencephalon is the ***subgranular zone*** in the hilus of the dentate gyrus that is the source of granule cells. The septum, fornix, and Ammon's horn have only a thin, darkly staining layer at the ventricle, and these are presumed to be generating glia, cells of the choroid plexus, and the ependymal lining of the ventricle.

Most of the structures in the diencephalon appear to be settled and are maturing, but the ***glioepithelium/ependyma*** lining the third ventricle is more thick than in the older specimens. A convoluted ***glioepithelium/ependyma*** lines the cerebral aqueduct in the midbrain. A smooth ***glioepithelium/ependyma*** lines the fourth ventricle through much of the pons. Another convoluted ***glioepithelium/ependyma*** lines the fourth ventricle through much of the medulla; part of that may be a remnant of the germinal source of the precerebellar nuclei. The ***external germinal layer*** is prominent over the entire surface of the cerebellar cortex and is actively producing basket, stellate, and granule cells. The ***germinal trigone*** is visible at the base of the nodulus and along the floccular peduncle; choroid plexus cells and glia are originating here and some migrating neurons may still be in the trigone.

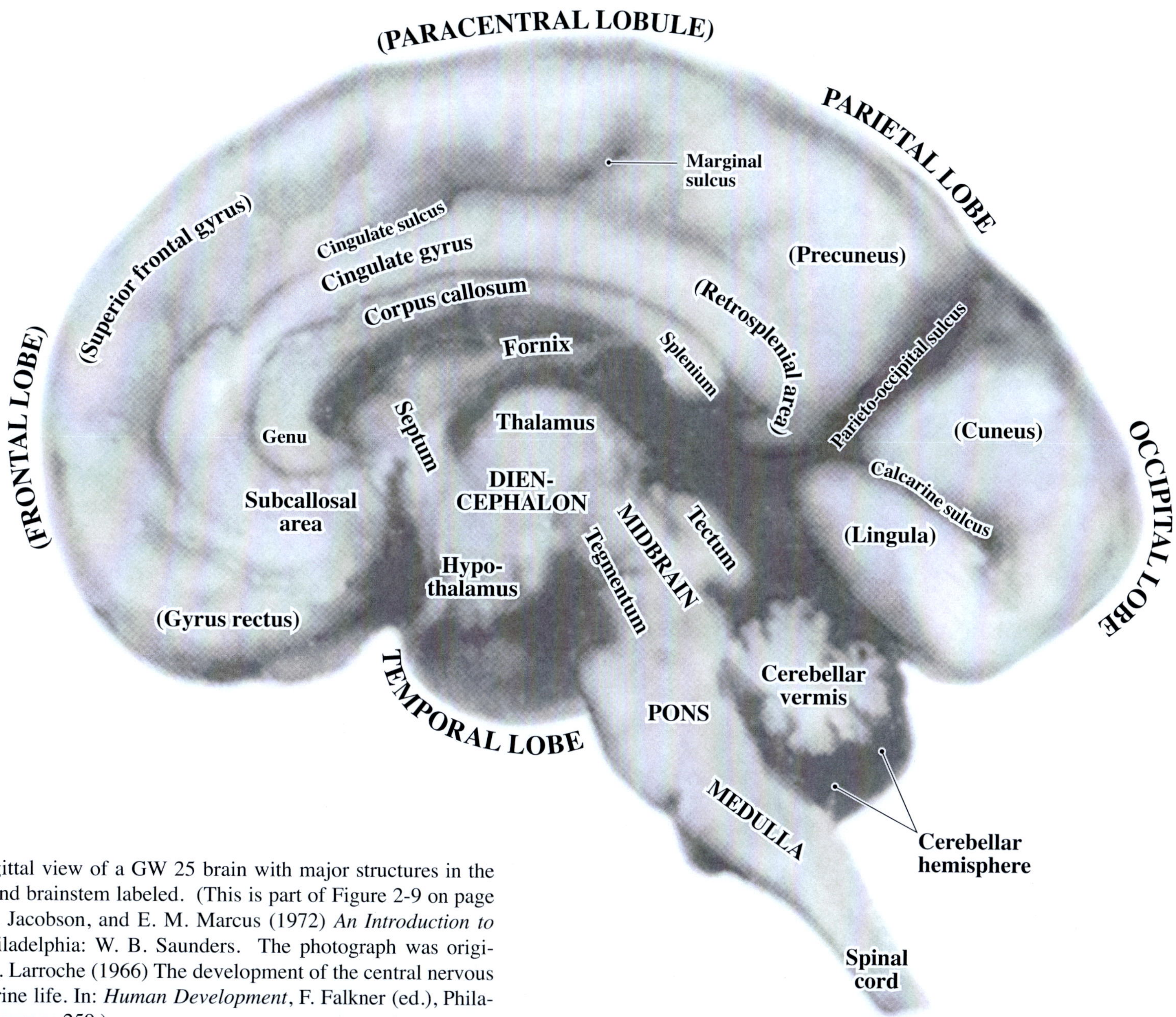

Figure 5. Midline sagittal view of a GW 25 brain with major structures in the cerebral hemispheres and brainstem labeled. (This is part of Figure 2-9 on page 27 in B. A. Curtis, S. Jacobson, and E. M. Marcus (1972) *An Introduction to the Neurosciences*, Philadelphia: W. B. Saunders. The photograph was originally published by J. C. Larroche (1966) The development of the central nervous system during intrauterine life. In: *Human Development*, F. Falkner (ed.), Philadelphia: W. B. Saunders, page 259.)

PLATE 1A
CR 225 mm
GW 26, Y147-63
Sagittal
Section 501

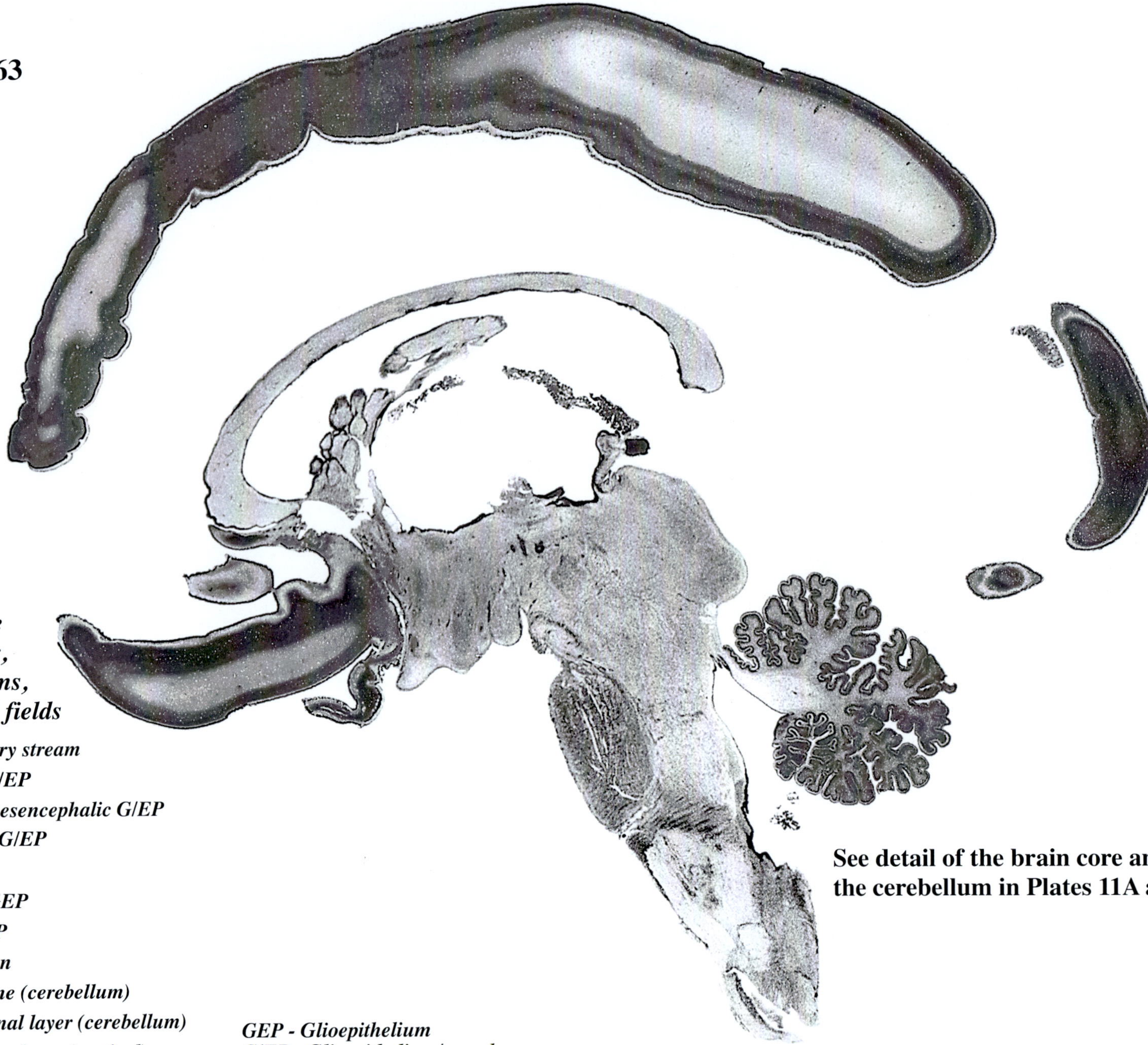

Remnants of the germinal matrix, migratory streams, and transitional fields

1 *Rostral migratory stream*
2 *Diencephalic G/EP*
3 *Diencephalic/mesencephalic G/EP*
4 *Mesencephalic G/EP*
5 *Pontine G/EP*
6 *Precerebellar GEP*
7 *Medullary G/EP*
8 *Raphe migration*
9 *Germinal trigone (cerebellum)*
10 *External germinal layer (cerebellum)*
11 *Subpial granular layer (cortical)*

GEP - Glioepithelium
G/EP - Glioepithelium/ependyma

See detail of the brain core and the cerebellum in Plates 11A and B.

PLATE 1B

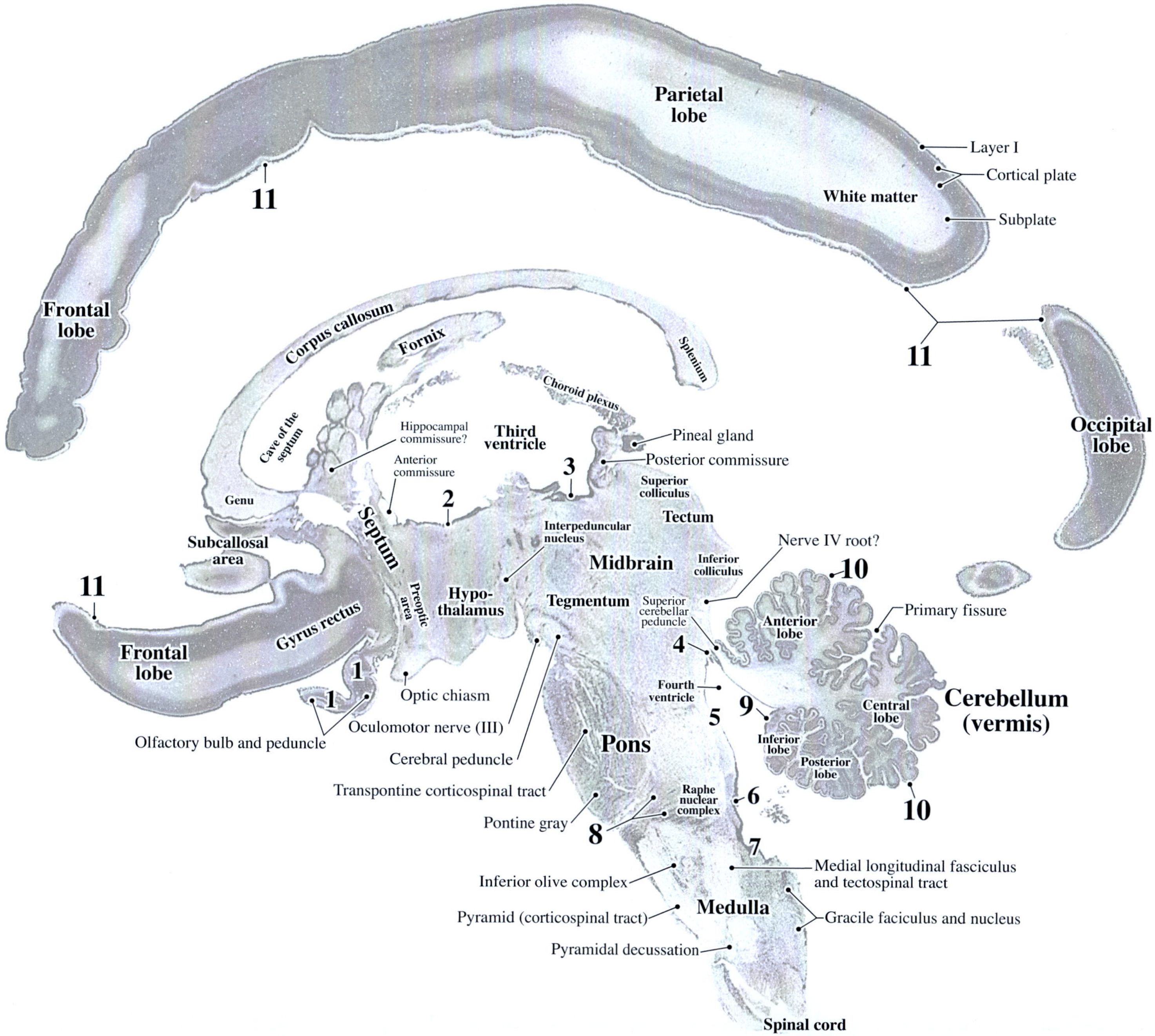

Parietal lobe
Layer I
Cortical plate
White matter
Subplate
11
Frontal lobe
Corpus callosum
Fornix
Splenium
Choroid plexus
Occipital lobe
Cave of the septum
Hippocampal commissure?
Anterior commissure
Third ventricle
Pineal gland
Posterior commissure
3
Superior colliculus
Genu
Septum
2
Tectum
Interpeduncular nucleus
Subcallosal area
Midbrain
Nerve IV root?
Inferior colliculus
10
Preoptic area
Hypo-thalamus
Tegmentum
Superior cerebellar peduncle
Primary fissure
Gyrus rectus
Anterior lobe
4
1
Optic chiasm
Fourth ventricle
Cerebellum (vermis)
9
Central lobe
Oculomotor nerve (III)
5
Olfactory bulb and peduncle
Pons
Inferior lobe
Cerebral peduncle
Posterior lobe
Transpontine corticospinal tract
Raphe nuclear complex
6
Pontine gray
8
7
Medial longitudinal fasciculus and tectospinal tract
Inferior olive complex
Medulla
Pyramid (corticospinal tract)
Gracile faciculus and nucleus
Pyramidal decussation
Spinal cord

PLATE 2A
CR 225 mm
GW 26, Y147-63
Sagittal
Section 481

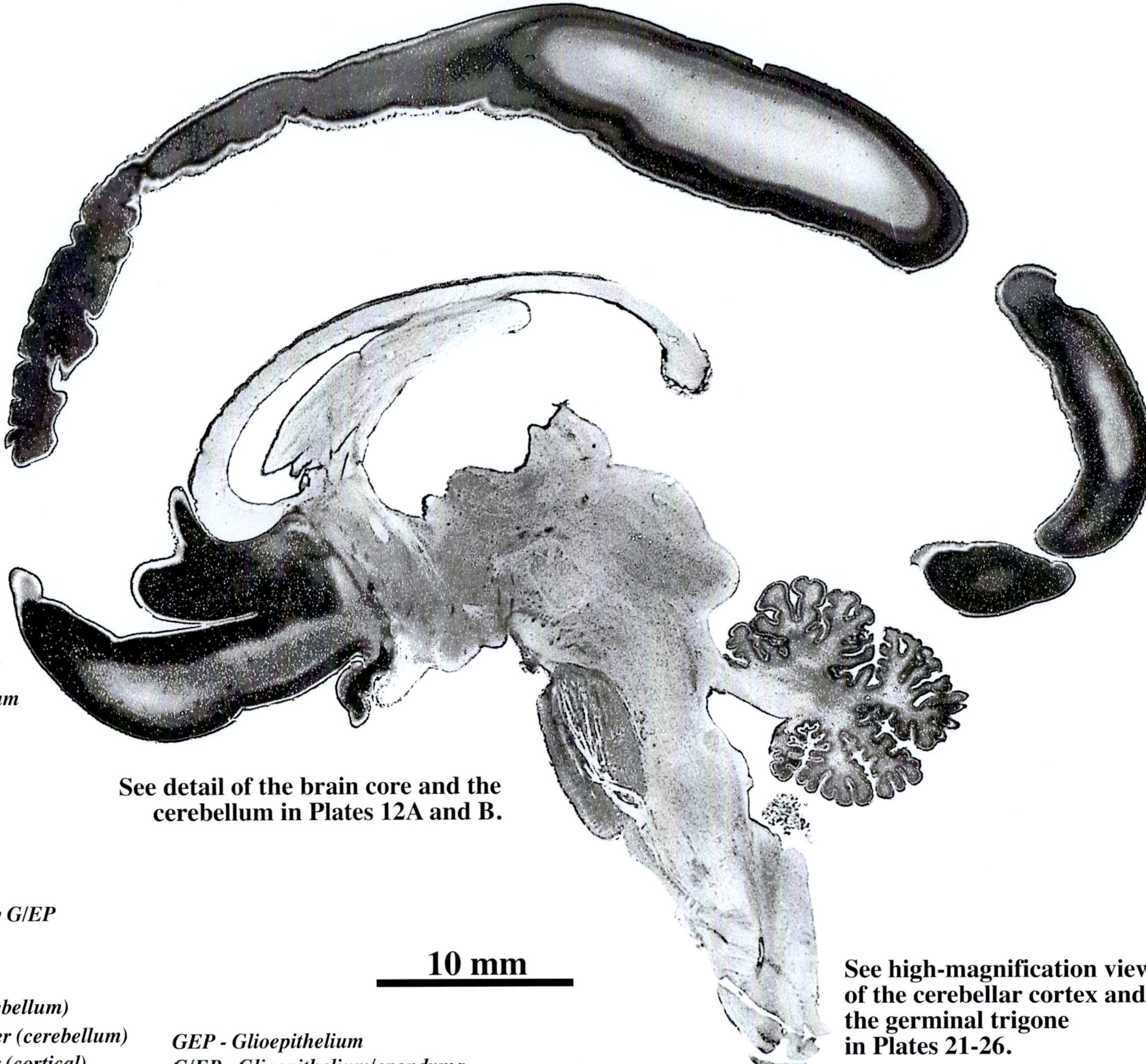

See detail of the brain core and the cerebellum in Plates 12A and B.

Remnants of the germinal matrix, migratory streams, and transitional fields

1 *Rostral migratory stream*
2 *Callosal sling*
3 *Callosal GEP*
4 *Fornical GEP*
5 *Strionuclear GEP*
6 *Thalamic G/EP*
7 *Mesencephalic G/EP*
8 *Pontine and medullary G/EP*
9 *Raphe migration*
10 *Spinal G/EP*
11 *Germinal trigone (cerebellum)*
12 *External germinal layer (cerebellum)*
13 *Subpial granular layer (cortical)*

GEP - Glioepithelium
G/EP - Glioepithelium/ependyma

See high-magnification views of the cerebellar cortex and the germinal trigone in Plates 21-26.

PLATE 2B

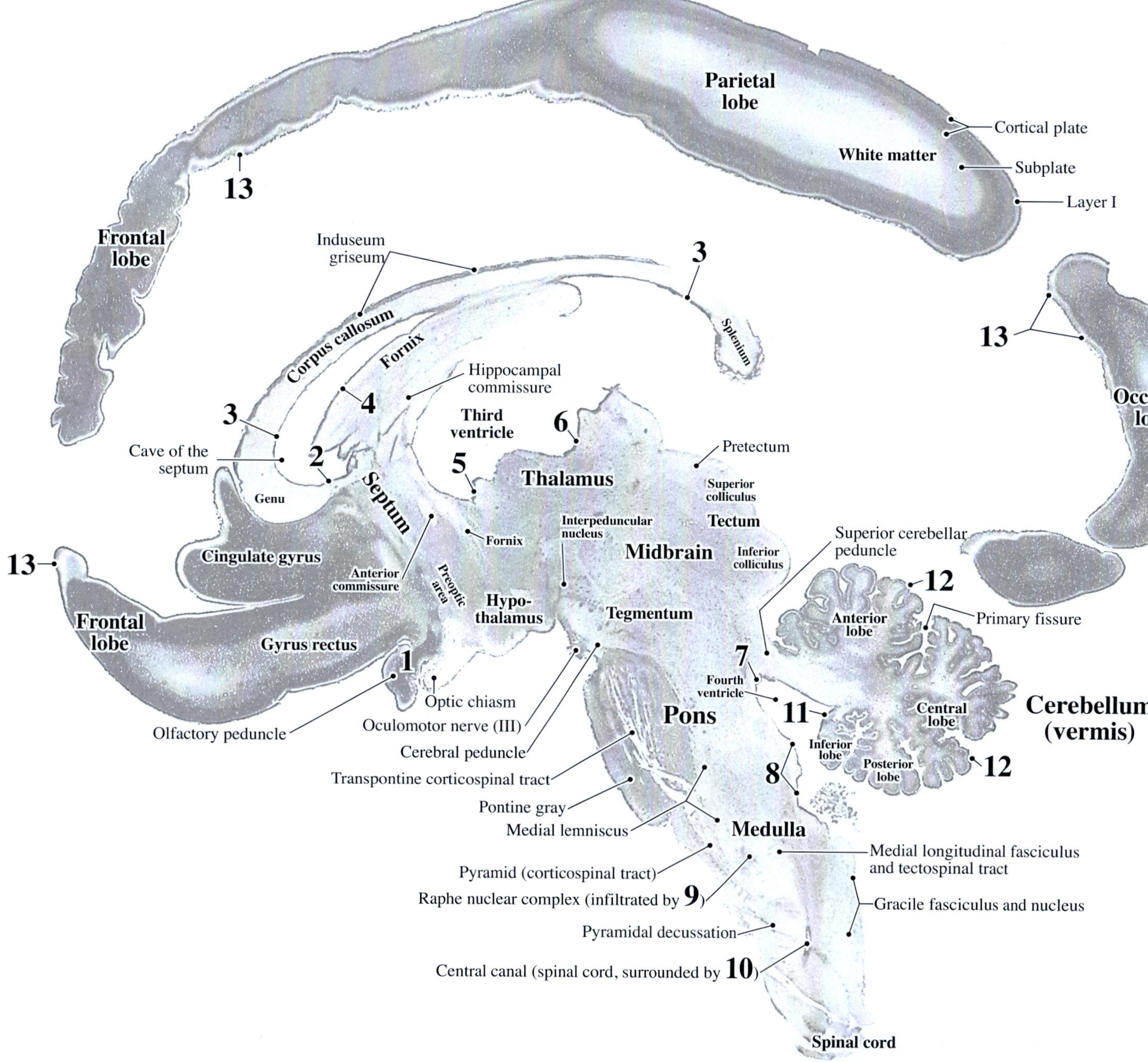
Parietal lobe
Cortical plate
White matter
Subplate
Layer I
13
Frontal lobe
Induseum griseum
3
Splenium
Corpus callosum
Fornix
Hippocampal commissure
13
Occipital lobe
3
4
Third ventricle
6
Cave of the septum
2
5
Pretectum
Thalamus
Superior colliculus
Genu
Septum
Tectum
Interpeduncular nucleus
Fornix
Superior cerebellar peduncle
Midbrain
Inferior colliculus
13
Cingulate gyrus
Anterior commissure
Preoptic area
12
Hypo-thalamus
Tegmentum
Anterior lobe
Primary fissure
Frontal lobe
Gyrus rectus
7
1
Fourth ventricle
Cerebellum (vermis)
Central lobe
Optic chiasm
11
Pons
Oculomotor nerve (III)
Olfactory peduncle
Inferior lobe
Cerebral peduncle
Posterior lobe
12
8
Transpontine corticospinal tract
Pontine gray
Medial lemniscus
Medulla
Medial longitudinal fasciculus and tectospinal tract
Pyramid (corticospinal tract)
Raphe nuclear complex (infiltrated by 9)
Gracile fasciculus and nucleus
Pyramidal decussation
Central canal (spinal cord, surrounded by 10)
Spinal cord

PLATE 3A
CR 225 mm
GW 26, Y147-63
Sagittal
Section 461

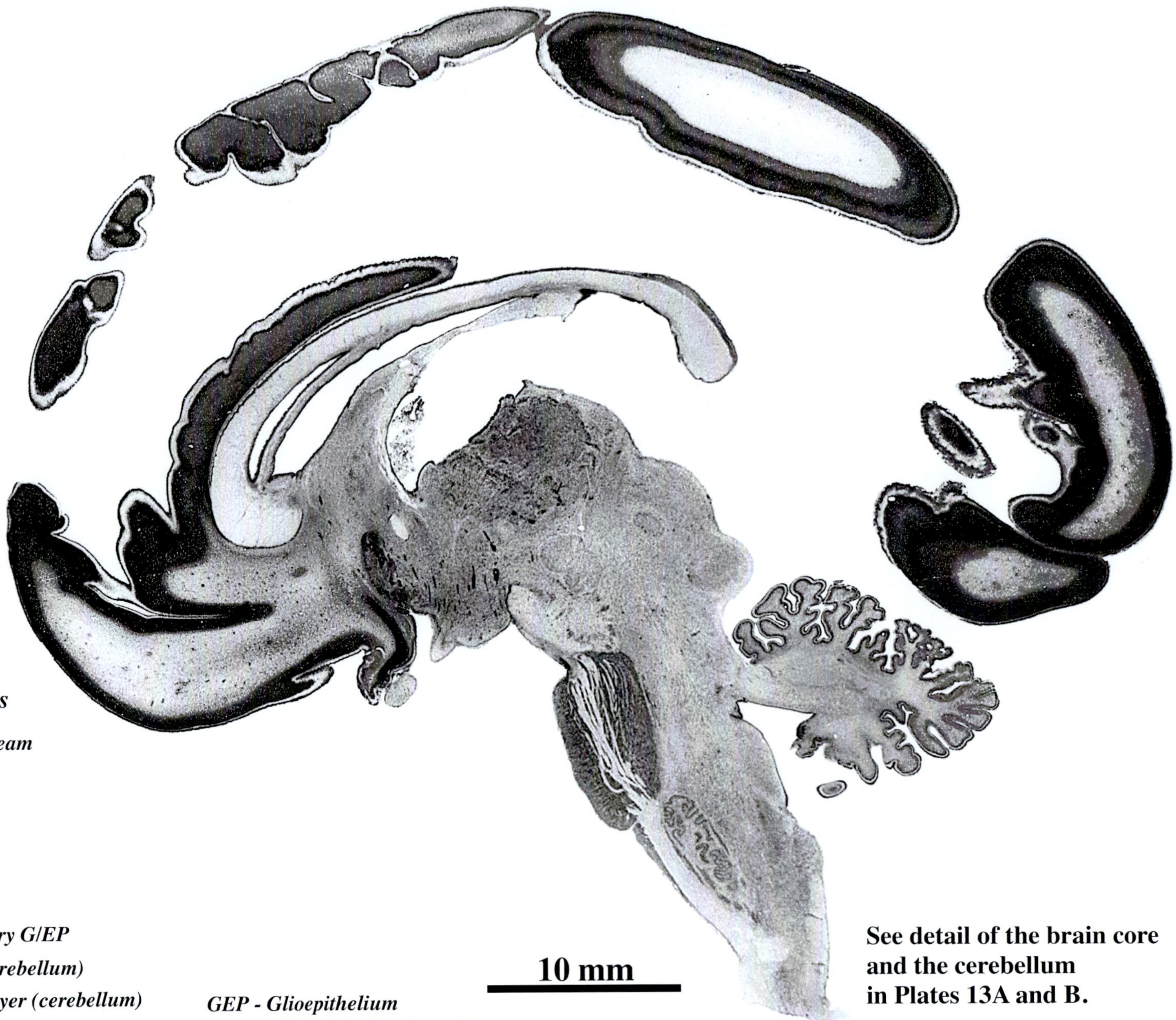

Remnants of the germinal matrix, migratory streams, and transitional fields

1 *Rostral migratory stream*
2 *Callosal GEP*
3 *Fornical GEP*
4 *Strionuclear GEP*
5 *Thalamic G/EP*
6 *Mesencephalic G/EP*
7 *Pontine and medullary G/EP*
8 *Germinal trigone (cerebellum)*
9 *External germinal layer (cerebellum)*
10 *Subpial granular layer (cortical)*

GEP - Glioepithelium
G/EP - Glioepithelium/ependyma

10 mm

See detail of the brain core and the cerebellum in Plates 13A and B.

PLATE 3B

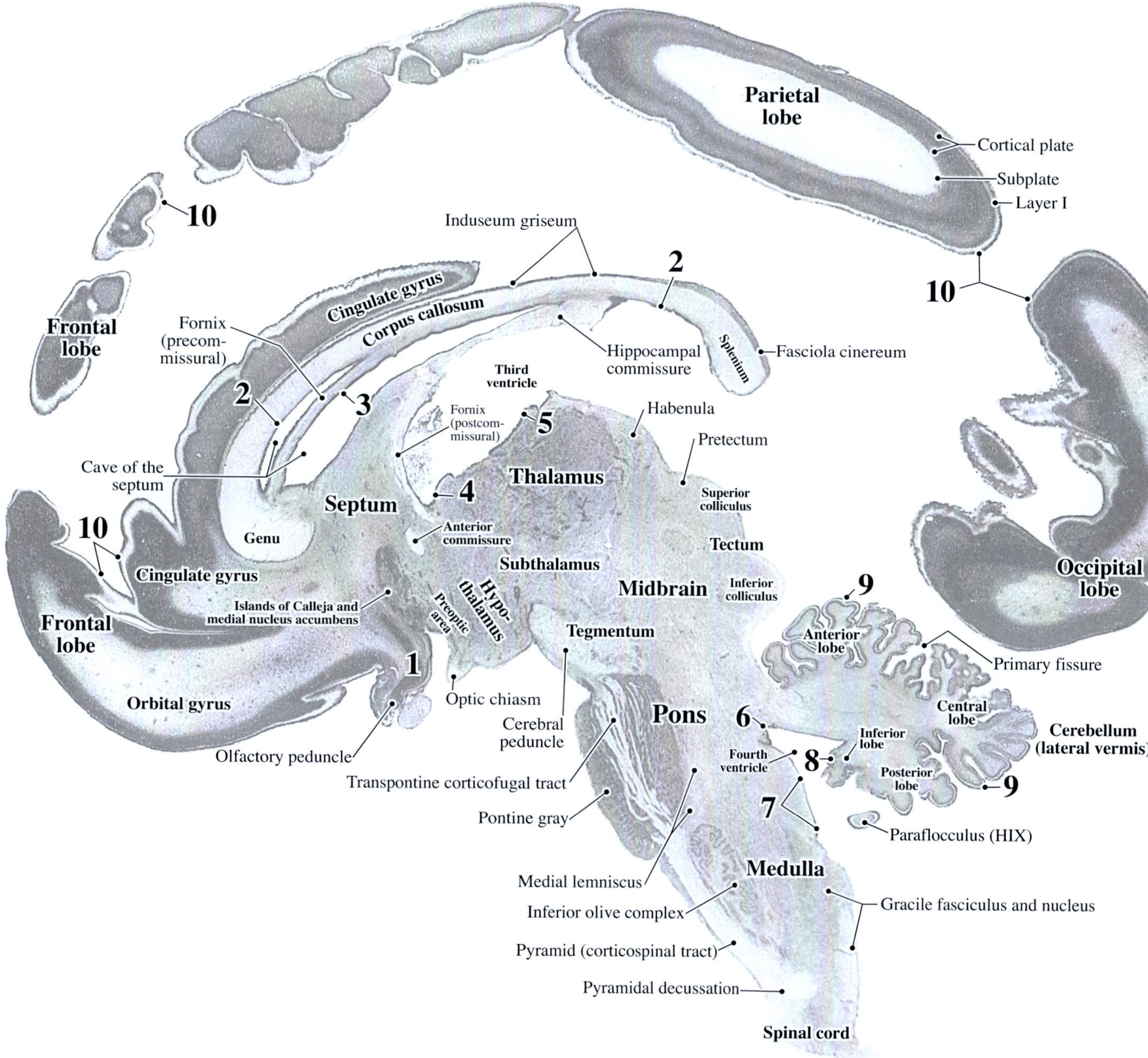
Parietal lobe
Cortical plate
Subplate
Layer I
10
Induseum griseum
2
10
Frontal lobe
Fornix (precom-missural)
Cingulate gyrus
Corpus callosum
Hippocampal commissure
Splenium
Fasciola cinereum
Third ventricle
2
3
Fornix (postcom-missural)
5
Habenula
Pretectum
Cave of the septum
Thalamus
4
Septum
Superior colliculus
10
Genu
Anterior commissure
Tectum
Subthalamus
Cingulate gyrus
Islands of Calleja and medial nucleus accumbens
Preoptic area
Hypo-thalamus
Midbrain
Inferior colliculus
9
Occipital lobe
Frontal lobe
Tegmentum
Anterior lobe
Primary fissure
1
Optic chiasm
Orbital gyrus
Pons
6
Central lobe
Cerebral peduncle
Inferior lobe
Cerebellum (lateral vermis)
Olfactory peduncle
Fourth ventricle
8
Posterior lobe
9
Transpontine corticofugal tract
7
Pontine gray
Paraflocculus (HIX)
Medulla
Medial lemniscus
Gracile fasciculus and nucleus
Inferior olive complex
Pyramid (corticospinal tract)
Pyramidal decussation
Spinal cord

PLATE 4A
CR 225 mm
GW 26, Y147-63
Sagittal
Section 421

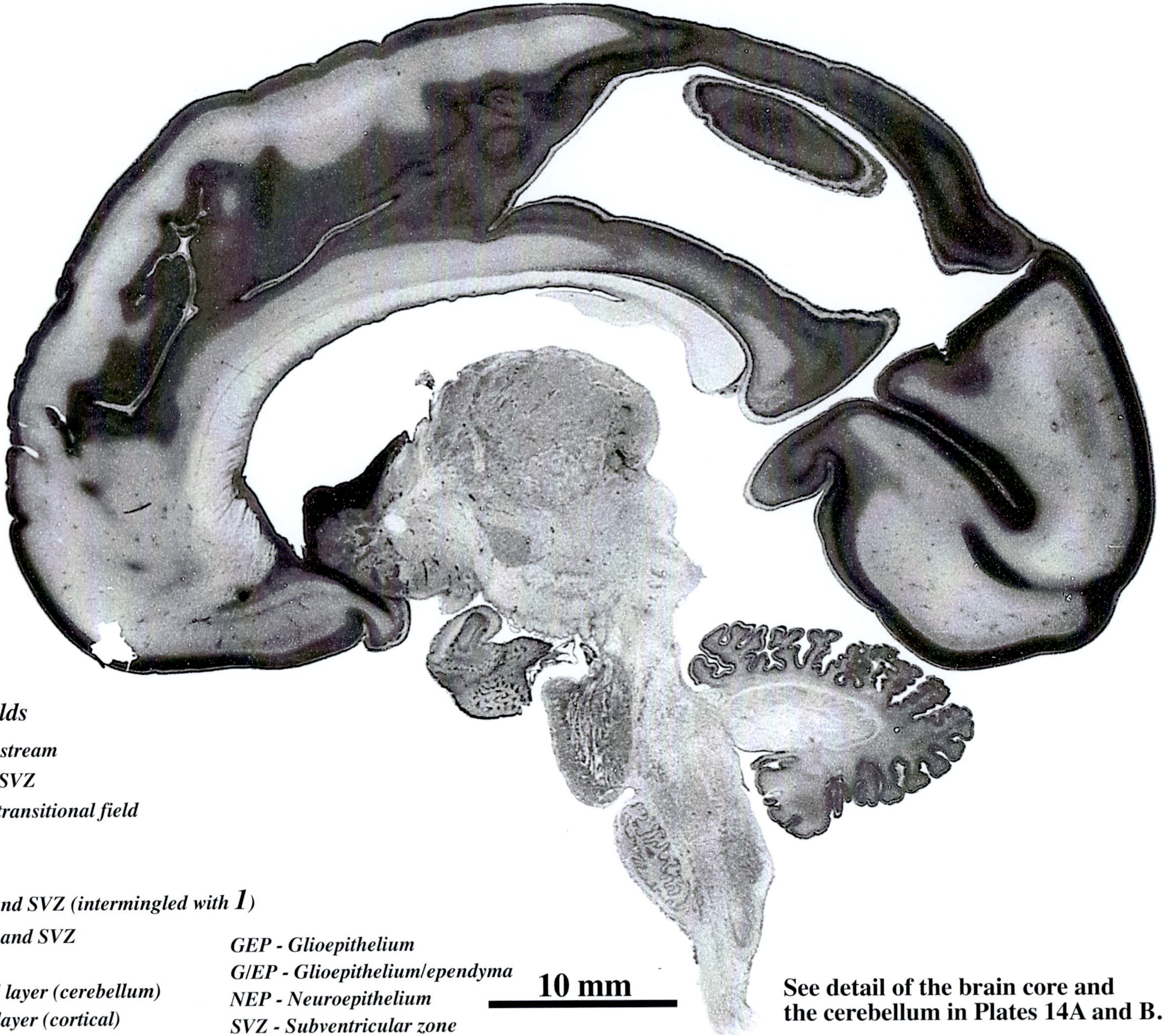

Remnants of the germinal matrix, migratory streams, and transitional fields

1 *Rostral migratory stream*
2 *Frontal NEP and SVZ*
3 *Frontal stratified transitional field*
4 *Callosal GEP*
5 *Fornical GEP*
6 *Accumbent NEP and SVZ (intermingled with 1)*
7 *Strionuclear NEP and SVZ*
8 *Pontine G/EP*
9 *External germinal layer (cerebellum)*
10 *Subpial granular layer (cortical)*

GEP - Glioepithelium
G/EP - Glioepithelium/ependyma
NEP - Neuroepithelium
SVZ - Subventricular zone

10 mm

See detail of the brain core and the cerebellum in Plates 14A and B.

PLATE 4B

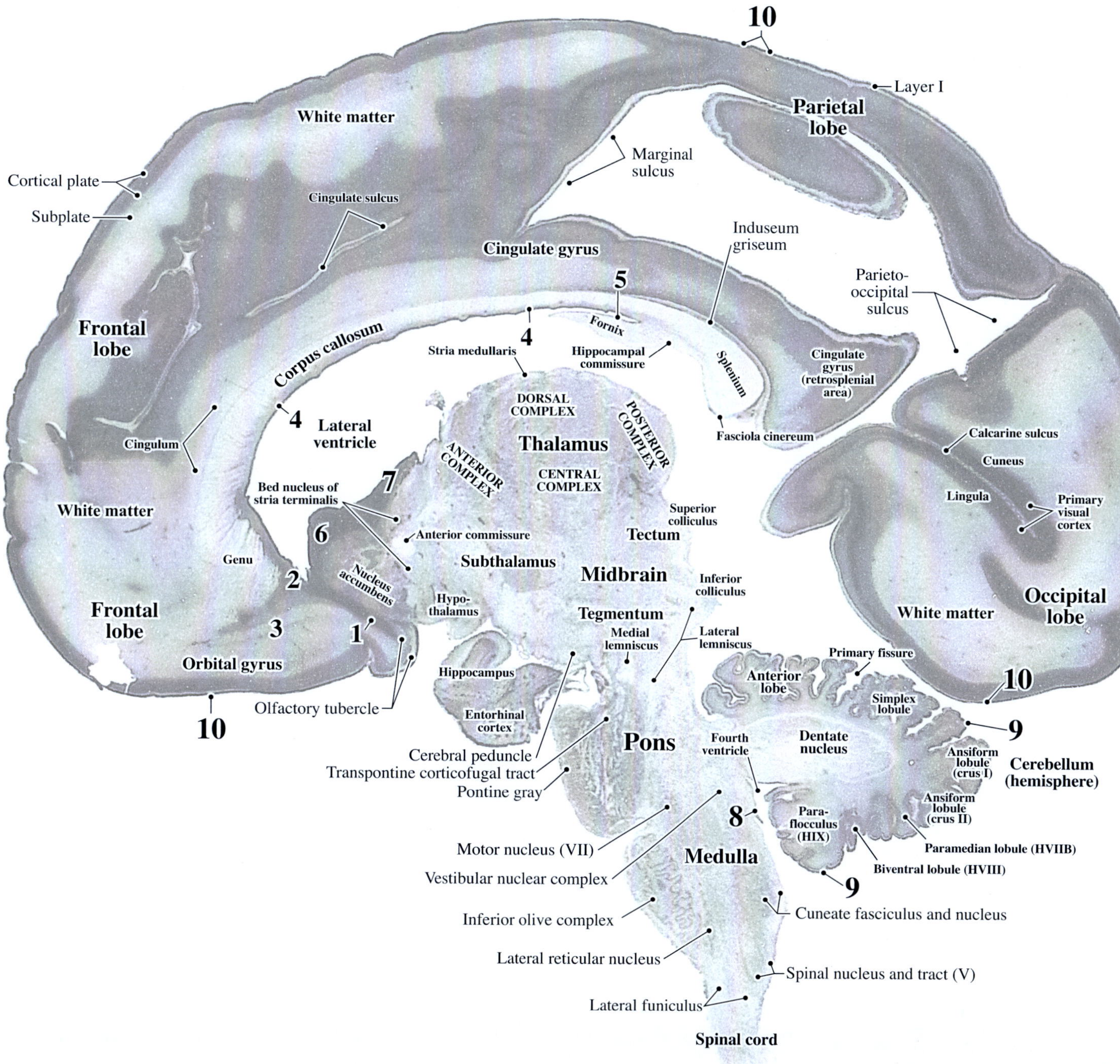
10
Layer I
Parietal lobe
White matter
Marginal sulcus
Cortical plate
Subplate
Cingulate sulcus
Induseum griseum
Cingulate gyrus
5
Parieto-occipital sulcus
Frontal lobe
Fornix
4
Corpus callosum
Stria medullaris
Hippocampal commissure
Splenium
Cingulate gyrus (retrosplenial area)
DORSAL COMPLEX
POSTERIOR COMPLEX
4
Lateral ventricle
Fasciola cinereum
Cingulum
Thalamus
Calcarine sulcus
ANTERIOR COMPLEX
Cuneus
CENTRAL COMPLEX
7
Bed nucleus of stria terminalis
Lingula
White matter
Superior colliculus
Primary visual cortex
6
Anterior commissure
Tectum
Genu
Subthalamus
Nucleus accumbens
2
Midbrain
Inferior colliculus
Frontal lobe
Hypo-thalamus
Tegmentum
White matter
Occipital lobe
3
1′
Medial lemniscus
Lateral lemniscus
Primary fissure
Orbital gyrus
Hippocampus
Anterior lobe
10
Simplex lobule
Olfactory tubercle
Entorhinal cortex
10
Pons
Fourth ventricle
Dentate nucleus
9
Ansiform lobule (crus I)
Cerebellum (hemisphere)
Cerebral peduncle
Transpontine corticofugal tract
Pontine gray
Ansiform lobule (crus II)
8
Para-flocculus (HIX)
Paramedian lobule (HVIIB)
Motor nucleus (VII)
Medulla
Biventral lobule (HVIII)
Vestibular nuclear complex
9
Cuneate fasciculus and nucleus
Inferior olive complex
Lateral reticular nucleus
Spinal nucleus and tract (V)
Lateral funiculus
Spinal cord

PLATE 5A
CR 225 mm
GW 26, Y147-63
Sagittal
Section 381

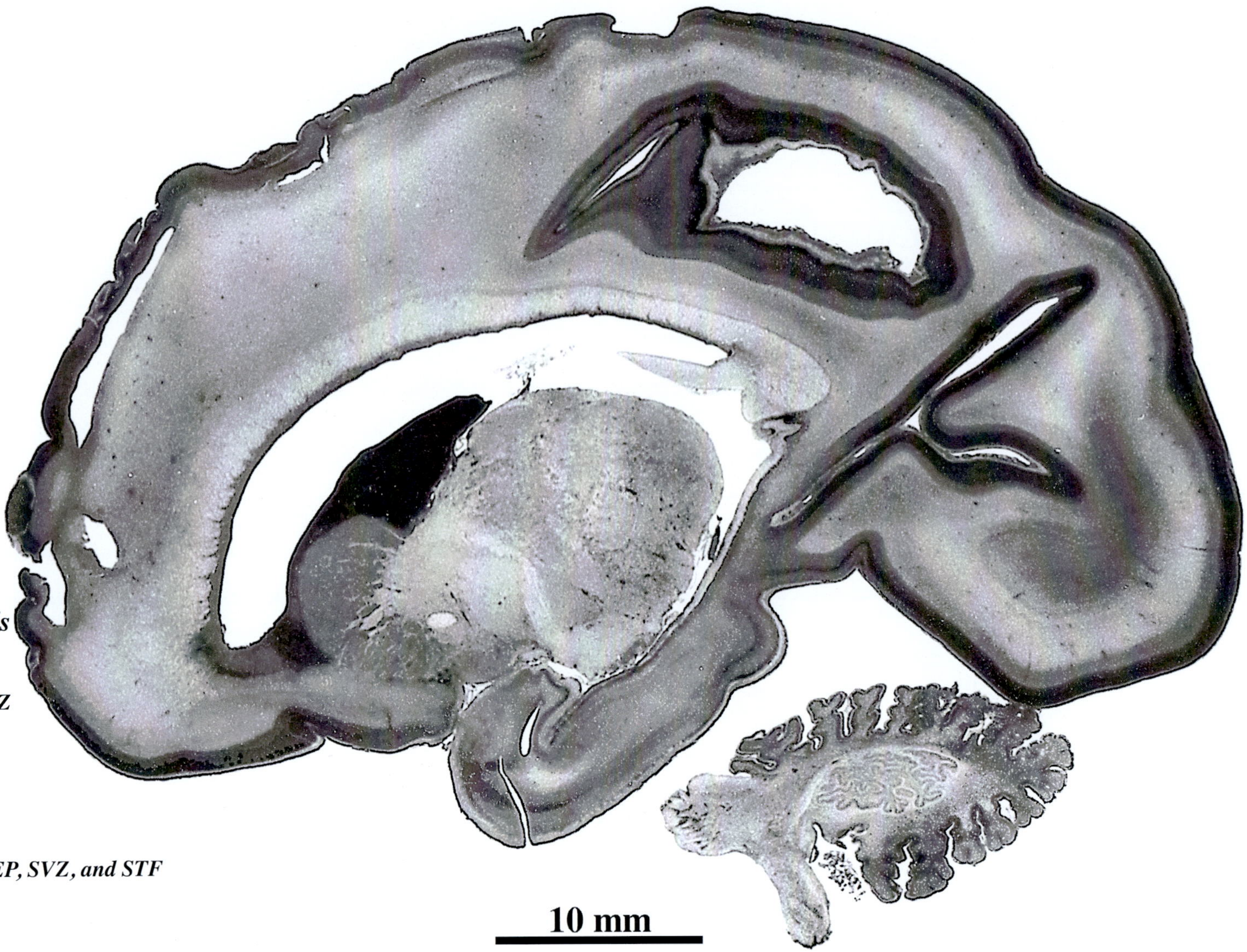

Remnants of the germinal matrix, migratory streams, and transitional fields

1 *Rostral migratory stream (source area)*
2 *Frontal NEP and SVZ*
3 *Frontal STF*
4 *Callosal GEP*
5 *Fornical GEP*
6 *Occipital STF*
7 *Parahippocampal NEP, SVZ, and STF*
8 *Alvear GEP*
9 *Amygdaloid GE/P*
10 *Anterolateral striatal NEP and SVZ*
11 *Anteromedial striatal NEP and SVZ*
12 *Strionuclear GEP*
13 *External germinal layer (cerebellum)*
14 *Subpial granular layer (cortical)*

GEP - Glioepithelium
G/EP - Glioepithelium/ependyma
NEP - Neuroepithelium
STF - Stratified transitional field
SVZ - Subventricular zone

See detail of the brain core and the cerebellum in Plates 15A and B.

PLATE 5B

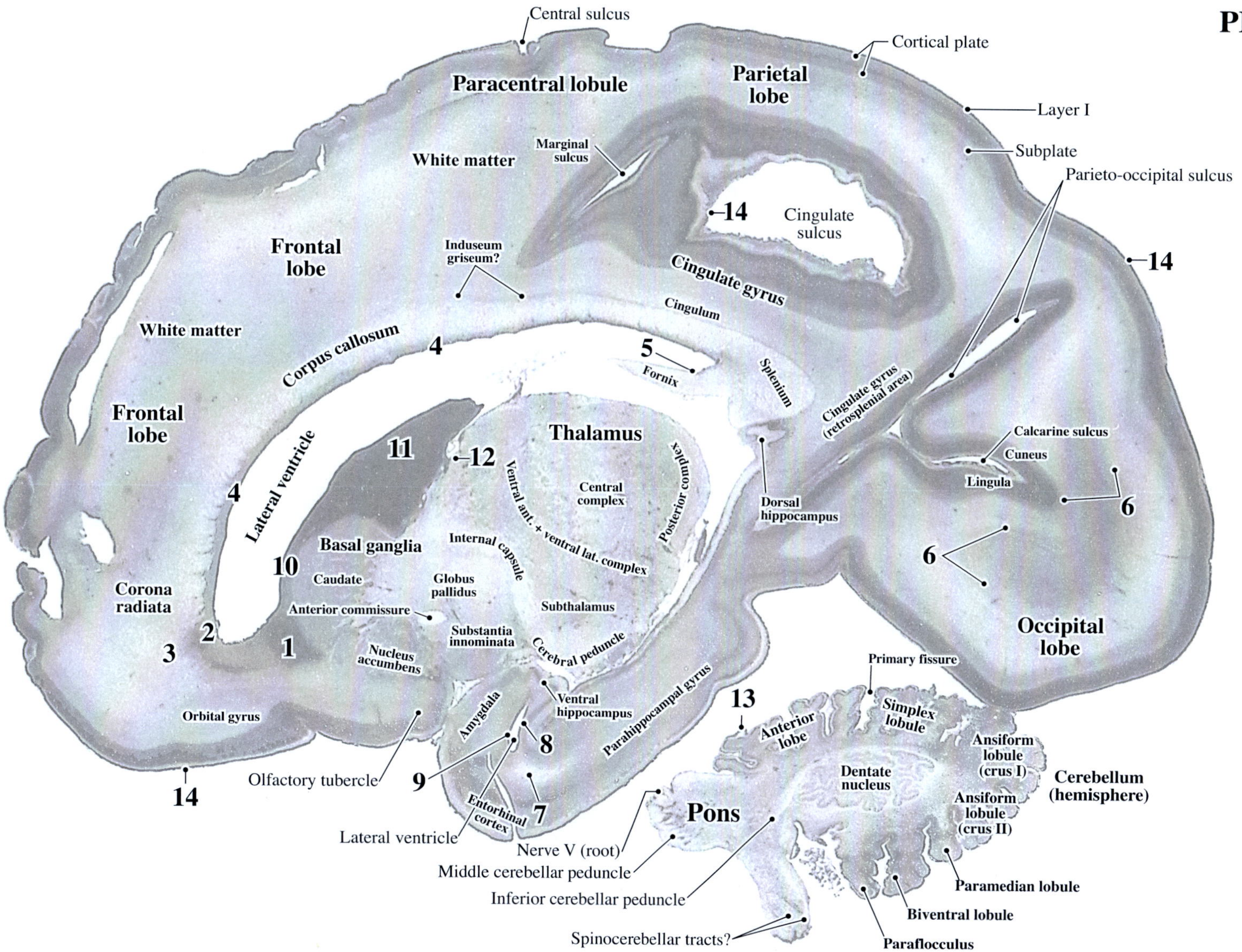
Central sulcus
Paracentral lobule
Parietal lobe
Cortical plate
Layer I
Subplate
Parieto-occipital sulcus
Marginal sulcus
White matter
14
Cingulate sulcus
14
Frontal lobe
Induseum griseum?
Cingulate gyrus
Cingulum
White matter
Corpus callosum
4
5
Fornix
Splenium
Cingulate gyrus (retrosplenial area)
Frontal lobe
Lateral ventricle
11
12
Thalamus
Calcarine sulcus
Cuneus
Lingula
6
Ventral ant. + ventral lat. complex
Central complex
Posterior complex
Dorsal hippocampus
4
Basal ganglia
Internal capsule
6
10
Caudate
Globus pallidus
Corona radiata
Anterior commissure
Subthalamus
Substantia innominata
Occipital lobe
2
3
1
Nucleus accumbens
Cerebral peduncle
Primary fissure
Ventral hippocampus
Amygdala
Parahippocampal gyrus
13
Simplex lobule
Anterior lobe
Orbital gyrus
8
Ansiform lobule (crus I)
Olfactory tubercle
9
Dentate nucleus
Cerebellum (hemisphere)
14
Entorhinal cortex
7
Pons
Ansiform lobule (crus II)
Lateral ventricle
Nerve V (root)
Middle cerebellar peduncle
Paramedian lobule
Inferior cerebellar peduncle
Biventral lobule
Spinocerebellar tracts?
Paraflocculus

PLATE 6A
CR 225 mm
GW 26, Y147-63
Sagittal
Section 361

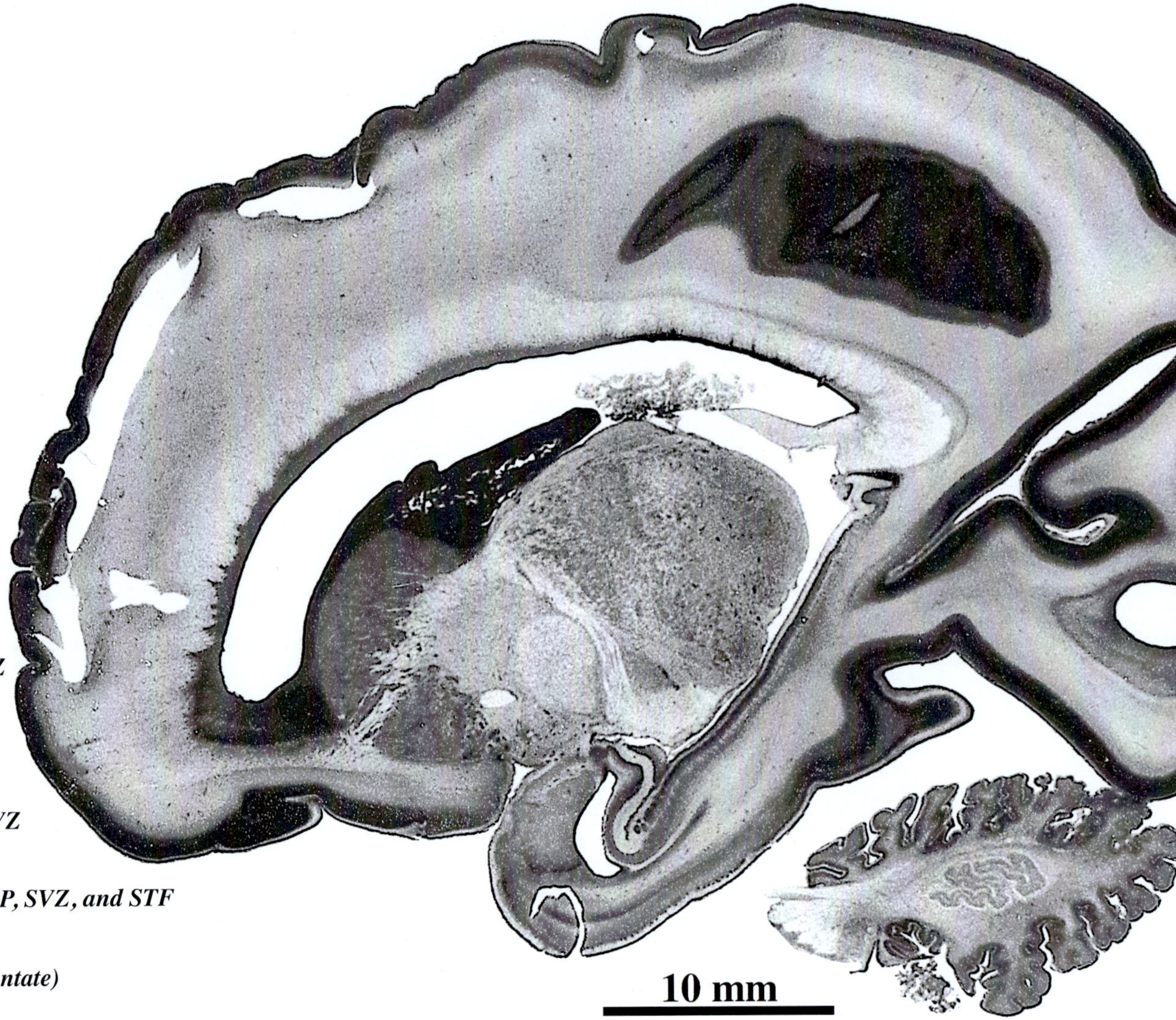

Remnants of the germinal matrix, migratory streams, and transitional fields

1 *Rostral migratory stream (source area)*
2 *Frontal NEP and SVZ*
3 *Frontal STF*
4 *Callosal GEP*
5 *Fornical GEP*
6 *Occipital NEP and SVZ*
7 *Occipital STF*
8 *Parahippocampal NEP, SVZ, and STF*
9 *Alvear GEP*
10 *Subgranular zone (dentate)*
11 *Amygdaloid G/EP*
12 *Anterolateral striatal NEP and SVZ*
13 *Anteromedial striatal NEP and SVZ*
14 *Strionuclear GEP*
15 *External germinal layer (cerebellum)*
16 *Subpial granular layer (cortical)*

GEP - Glioepithelium
G/EP - Glioepithelium/ependyma
NEP - Neuroepithelium
STF - Stratified transitional field
SVZ - Subventricular zone

See detail of the brain core and the cerebellum in Plates 16A and B.

PLATE 6B

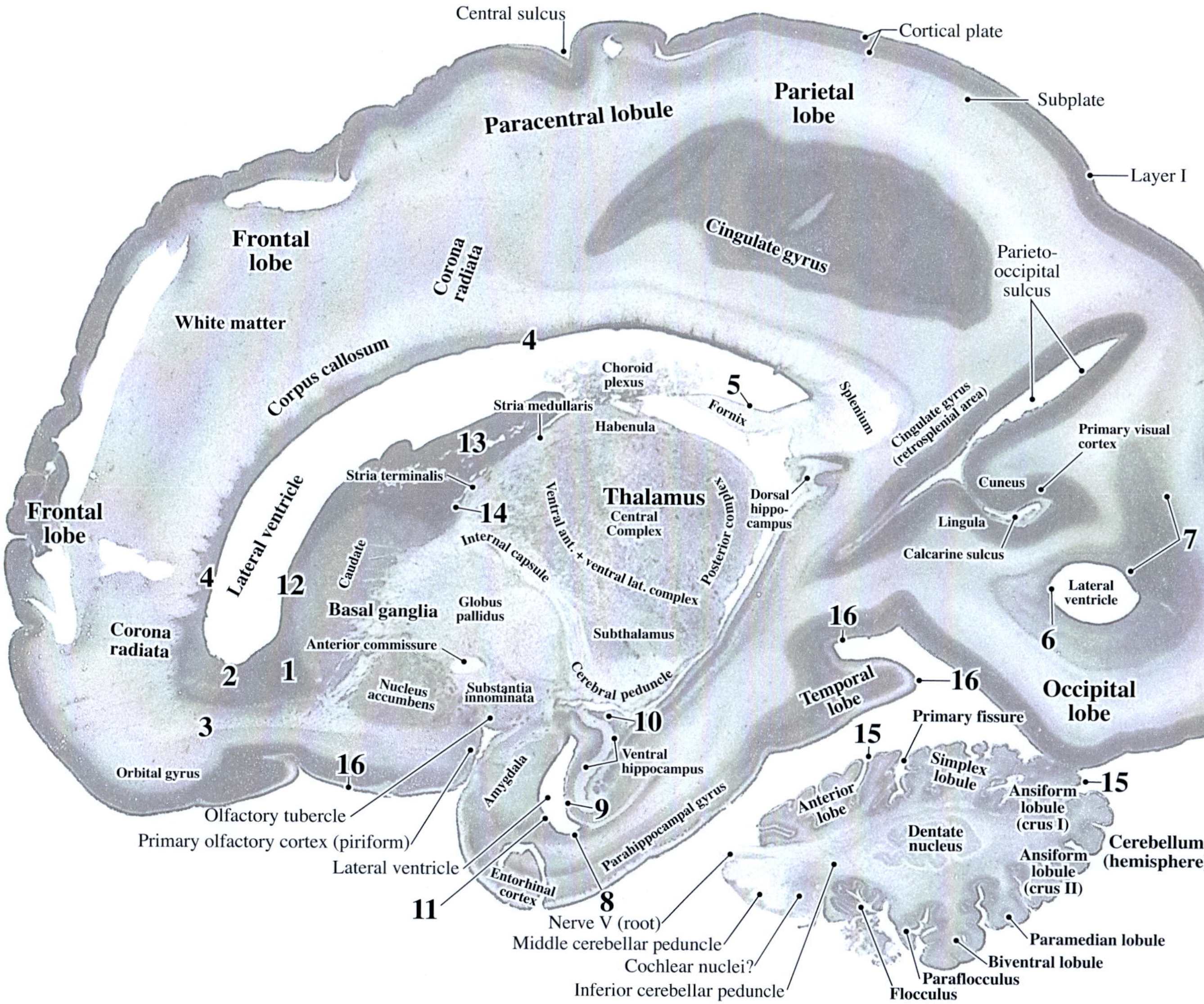
Central sulcus
Cortical plate
Subplate
Layer I
Paracentral lobule
Parietal lobe
Frontal lobe
Corona radiata
Cingulate gyrus
Parieto-occipital sulcus
16
White matter
Corpus callosum
4
Choroid plexus
5
Fornix
Splenium
Cingulate gyrus (retrosplenial area)
Stria medullaris
Habenula
13
Primary visual cortex
Stria terminalis
14
Thalamus
Central Complex
Ventral ant. + ventral lat. complex
Posterior complex
Dorsal hippocampus
Cuneus
Lingula
Calcarine sulcus
7
Frontal lobe
Lateral ventricle
Internal capsule
4
12
Caudate
Basal ganglia
Globus pallidus
Lateral ventricle
6
Corona radiata
Subthalamus
16
Anterior commissure
2
1
Nucleus accumbens
Substantia innominata
Cerebral peduncle
Temporal lobe
16
Occipital lobe
3
10
Primary fissure
Ventral hippocampus
15
Orbital gyrus
16
Amygdala
Simplex lobule
Anterior lobe
Ansiform lobule (crus I)
15
9
Parahippocampal gyrus
Dentate nucleus
Cerebellum (hemisphere)
Olfactory tubercle
Primary olfactory cortex (piriform)
Lateral ventricle
Entorhinal cortex
Ansiform lobule (crus II)
11
8
Nerve V (root)
Middle cerebellar peduncle
Cochlear nuclei?
Inferior cerebellar peduncle
Paramedian lobule
Biventral lobule
Paraflocculus
Flocculus

PLATE 7A
CR 225 mm
GW 26, Y147-63
Sagittal
Section 341

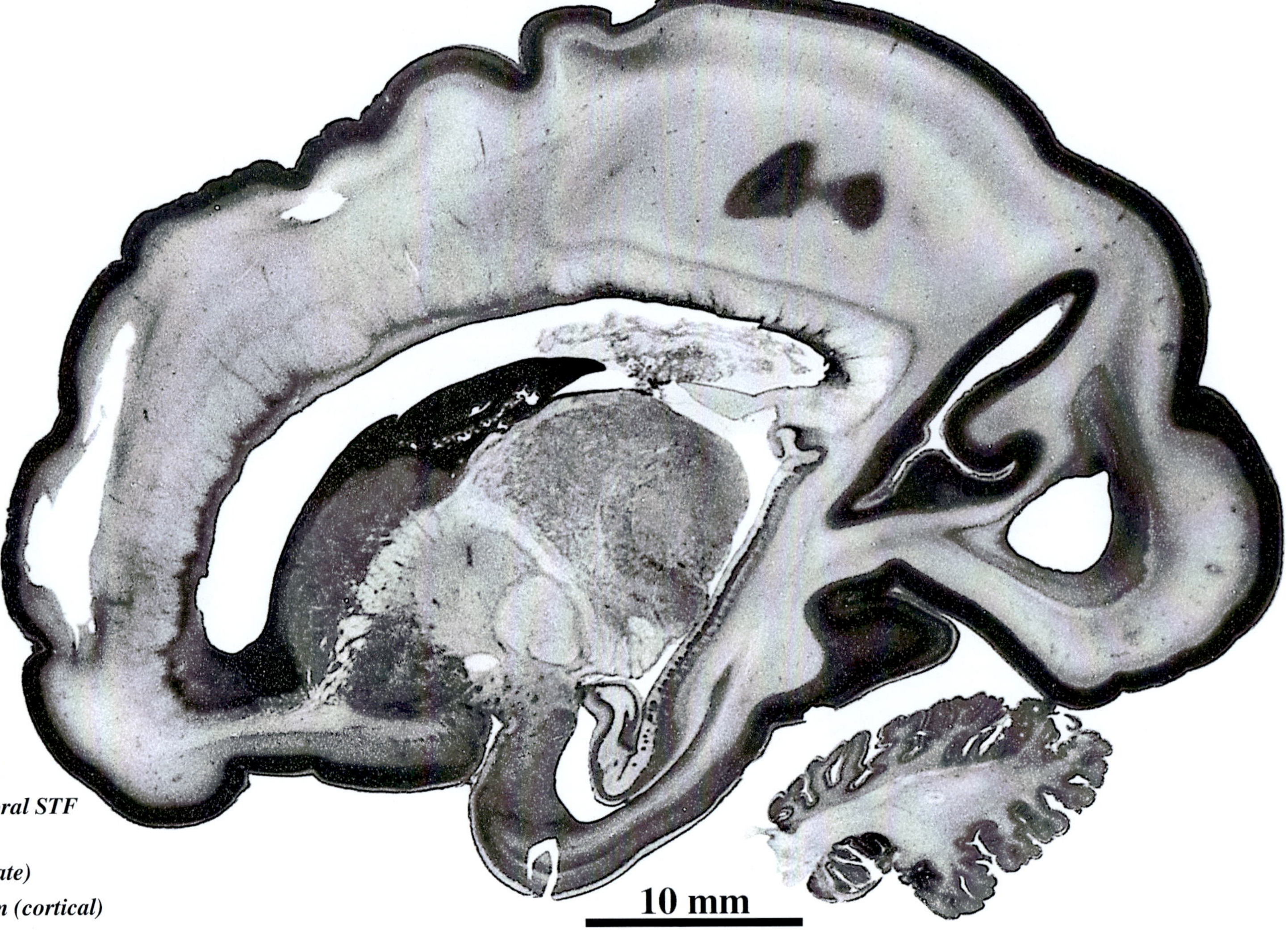

Remnants of the germinal matrix, migratory streams, and transitional fields

1. *Rostral migratory stream (source area)*
2. *Frontal NEP and SVZ*
3. *Frontal STF*
4. *Callosal GEP*
5. *Fornical GEP*
6. *Occipital NEP and SVZ*
7. *Occipital STF*
8. *Temporal NEP and SVZ*
9. *Parahippocampal/temporal STF*
10. *Alvear GEP*
11. *Subgranular zone (dentate)*
12. *Lateral migratory stream (cortical)*
13. *Amygdaloid G/EP*
14. *Anterolateral striatal NEP and SVZ*
15. *Anteromedial striatal NEP and SVZ*
16. *Strionuclear GEP*
17. *External germinal layer (cerebellum)*
18. *Subpial granular layer (cortical)*

GEP - Glioepithelium
G/EP - Glioepithelium/ependyma
NEP - Neuroepithelium
STF - Stratified transitional field
SVZ - Subventricular zone

See detail of the brain core and the cerebellum in Plates 17A and B.

PLATE 7B

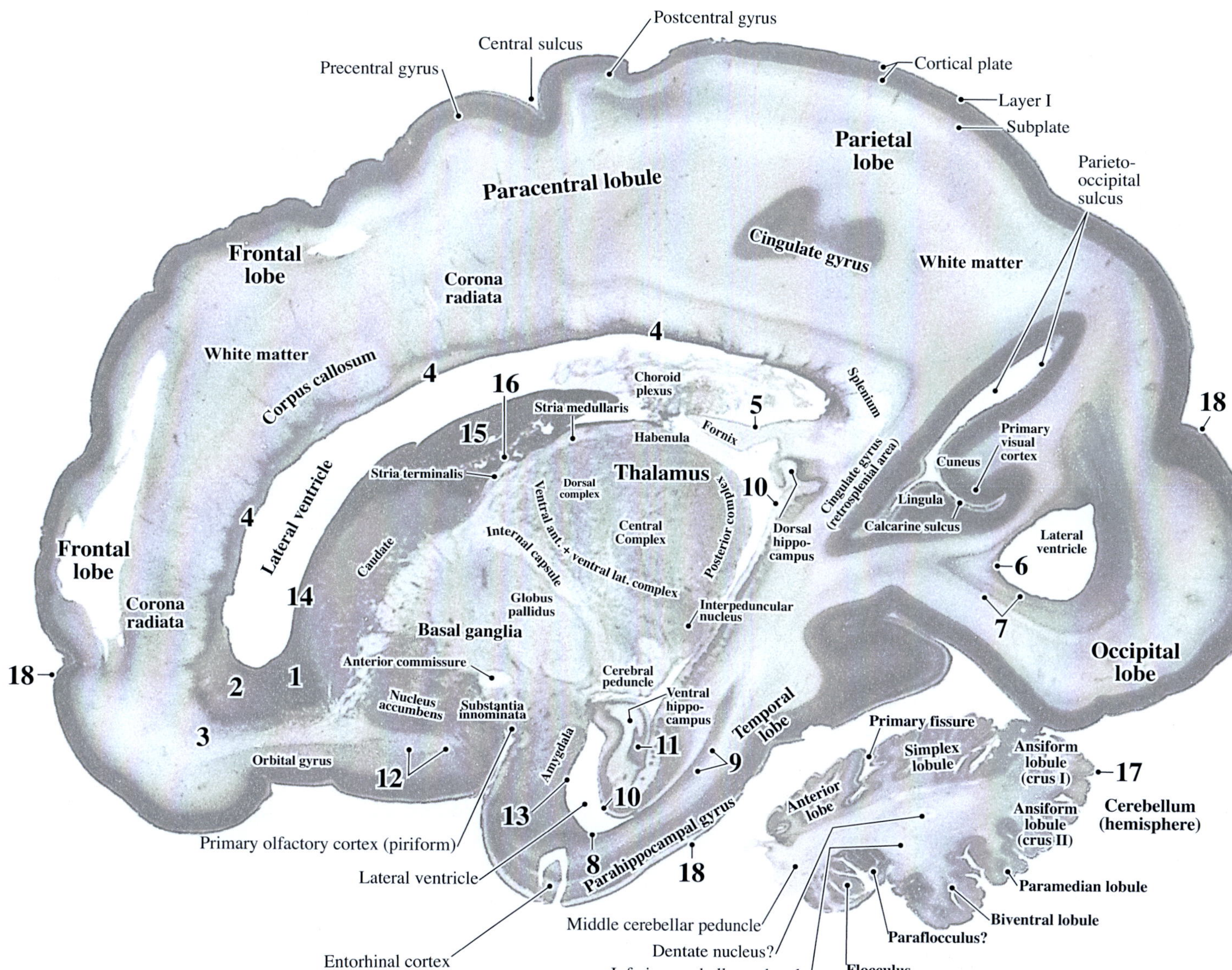
Postcentral gyrus
Central sulcus
Precentral gyrus
Cortical plate
Layer I
Subplate
Parietal lobe
Parieto-occipital sulcus
Paracentral lobule
Cingulate gyrus
White matter
Frontal lobe
Corona radiata
White matter
Corpus callosum
4
4
4
16
Choroid plexus
Splenium
18
5
Stria medullaris
Fornix
Primary visual cortex
15
Habenula
Stria terminalis
Thalamus
Cuneus
Cingulate gyrus (retrosplenial area)
10
Dorsal complex
Lingula
Posterior complex
Ventral ant. + ventral lat. complex
Dorsal hippocampus
Calcarine sulcus
Lateral ventricle
Lateral ventricle
Central Complex
Frontal lobe
Caudate
Internal capsule
6
Corona radiata
14
Globus pallidus
Interpeduncular nucleus
7
Basal ganglia
Occipital lobe
18
1
2
Anterior commissure
Cerebral peduncle
Ventral hippocampus
Nucleus accumbens
Substantia innominata
Temporal lobe
Primary fissure
3
11
Simplex lobule
Ansiform lobule (crus I)
17
Orbital gyrus
Amygdala
9
12
10
Cerebellum (hemisphere)
Anterior lobe
Ansiform lobule (crus II)
13
Primary olfactory cortex (piriform)
8
Parahippocampal gyrus
18
Lateral ventricle
Paramedian lobule
Biventral lobule
Middle cerebellar peduncle
Paraflocculus?
Entorhinal cortex
Dentate nucleus?
Inferior cerebellar peduncle
Flocculus

PLATE 8A
CR 225 mm
GW 26, Y147-63
Sagittal
Section 321

Remnants of the germinal matrix, migratory streams, and transitional fields

1. *Rostral migratory stream (source area)*
2. *Frontal NEP and SVZ*
3. *Frontal STF*
4. *Paracentral NEP and SVZ*
5. *Paracentral STF*
6. *Callosal GEP*
7. *Fornical GEP*
8. *Occipital NEP and SVZ*
9. *Occipital STF*
10. *Temporal NEP and SVZ*
11. *Temporal STF*
12. *Alvear GEP*
13. *Subgranular zone (dentate)*
14. *Lateral migratory stream (cortical)*
15. *Amygdaloid G/EP*
16. *Anterolateral striatal NEP and SVZ*
17. *Anteromedial/posterior striatal NEP and SVZ*
18. *Strionuclear GEP*
19. *External germinal layer (cerebellum)*
20. *Subpial granular layer (cortical)*

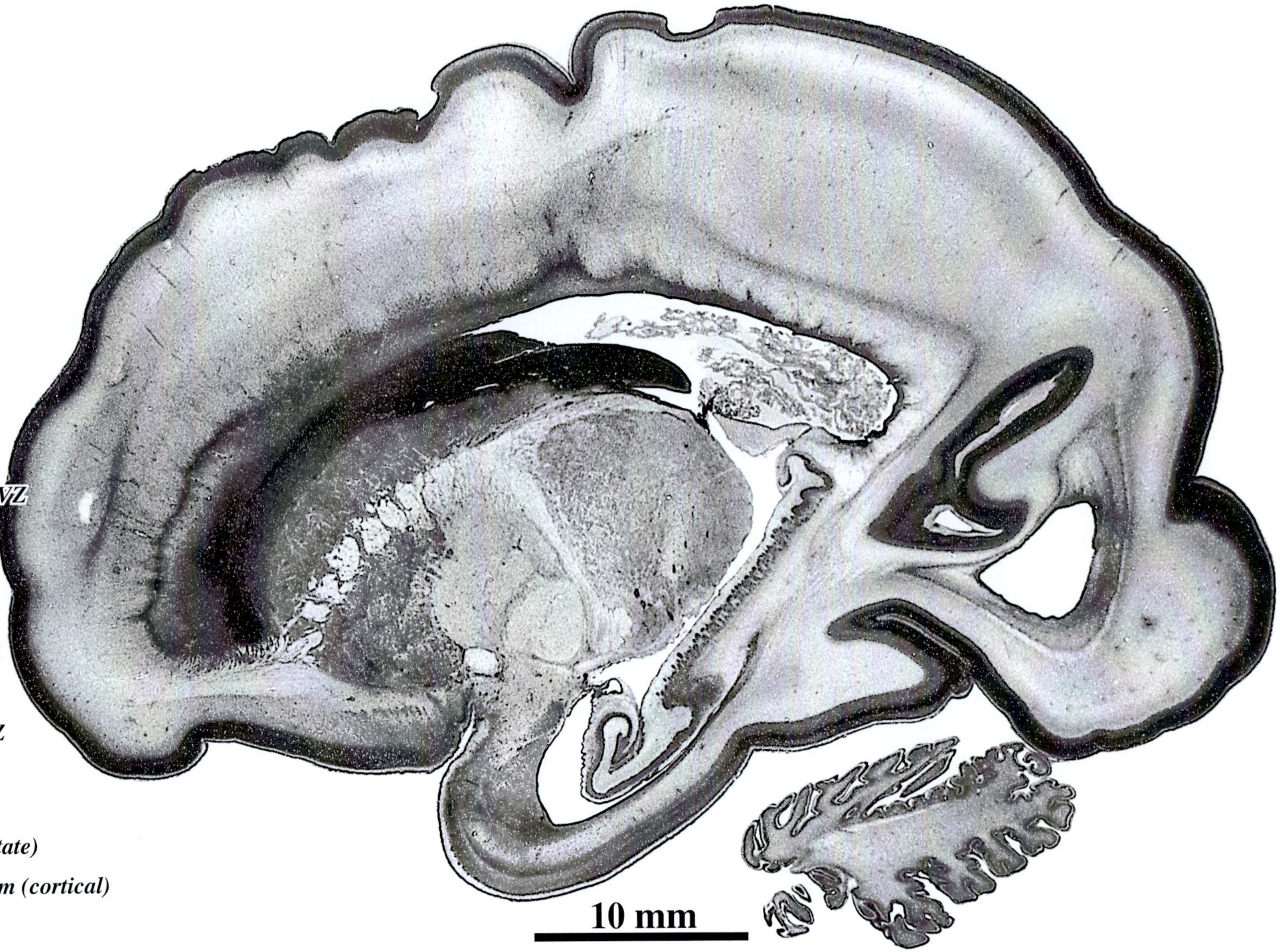

GEP - Glioepithelium
G/EP - Glioepithelium/ependyma
NEP - Neuroepithelium
STF - Stratified transitional field
SVZ - Subventricular zone

See detail of the brain core and the cerebellum in Plates 18A and B.

PLATE 8B

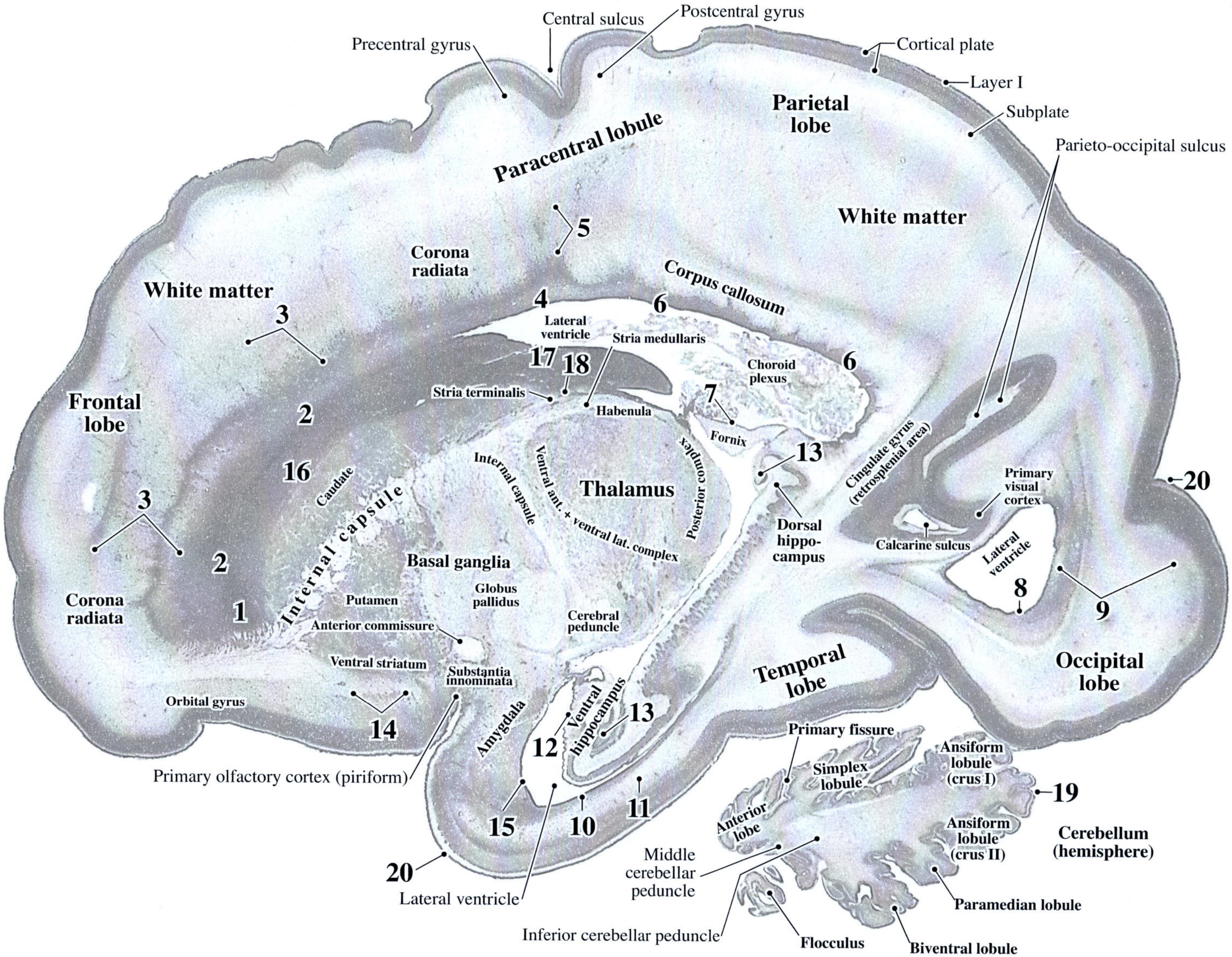
Central sulcus
Postcentral gyrus
Precentral gyrus
Cortical plate
Layer I
Parietal lobe
Subplate
Paracentral lobule
Parieto-occipital sulcus
White matter
5
Corona radiata
Corpus callosum
White matter
4
3
6
Lateral ventricle
Stria medullaris
17
18
Choroid plexus
6
Frontal lobe
Stria terminalis
7
2
Habenula
Fornix
13
Cingulate gyrus (retrosplenial area)
16
Internal capsule
Ventral ant. + ventral lat. complex
Thalamus
Posterior complex
Primary visual cortex
20
Caudate
3
Internal capsule
Dorsal hippocampus
Calcarine sulcus
2
Basal ganglia
Lateral ventricle
Globus pallidus
8
9
Corona radiata
1
Putamen
Cerebral peduncle
Anterior commissure
Temporal lobe
Occipital lobe
Ventral striatum
Substantia innominata
Orbital gyrus
Ventral hippocampus
13
14
Amygdala
12
Primary fissure
Ansiform lobule (crus I)
Simplex lobule
Primary olfactory cortex (piriform)
19
15
10
11
Anterior lobe
Ansiform lobule (crus II)
Cerebellum (hemisphere)
20
Middle cerebellar peduncle
Lateral ventricle
Paramedian lobule
Inferior cerebellar peduncle
Flocculus
Biventral lobule

PLATE 9A
CR 225 mm
GW 26, Y147-63
Sagittal
Section 261

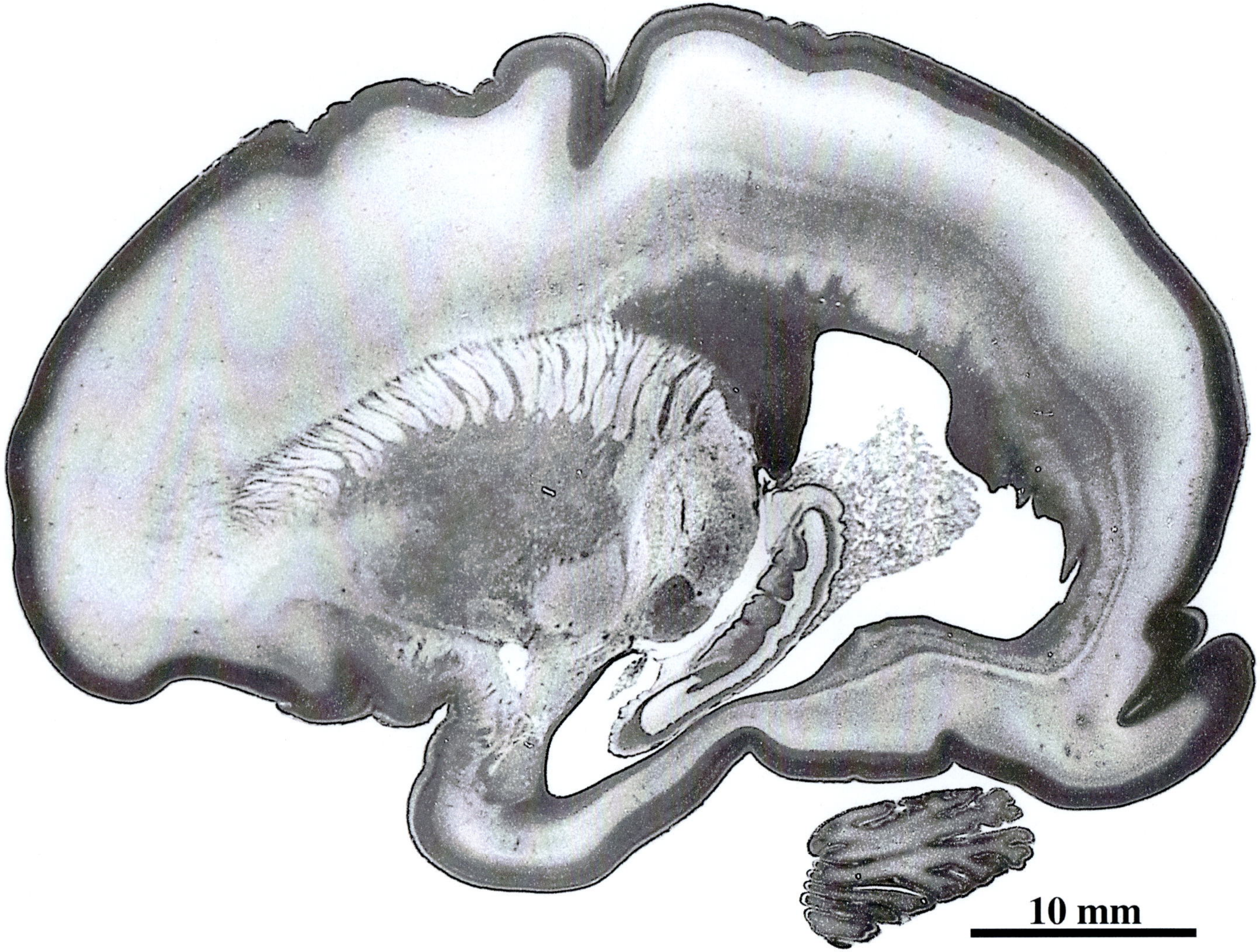

Remnants of the germinal matrix, migratory streams, and transitional fields

1 *Frontal STF*
2 *Paracentral STF*
3 *Parietal NEP and SVZ*
4 *Parietal STF*
5 *Occipital NEP and SVZ*
6 *Occipital STF*
7 *Temporal NEP and SVZ*
8 *Temporal STF*
9 *Alvear GEP*
10 *Subgranular zone (dentate)*
11 *Lateral migratory stream (cortical)*
12 *Amygdaloid G/EP*
13 *Posterior striatal NEP and SVZ*
14 *Strionuclear GEP*
15 *External germinal layer (cerebellum)*
16 *Subpial granular layer (cortical)*

GEP - Glioepithelium
G/EP - Glioepithelium/ependyma
NEP - Neuroepithelium
STF - Stratified transitional field
SVZ - Subventricular zone

See detail of the brain core in Plates 19A and B.

PLATE 9B

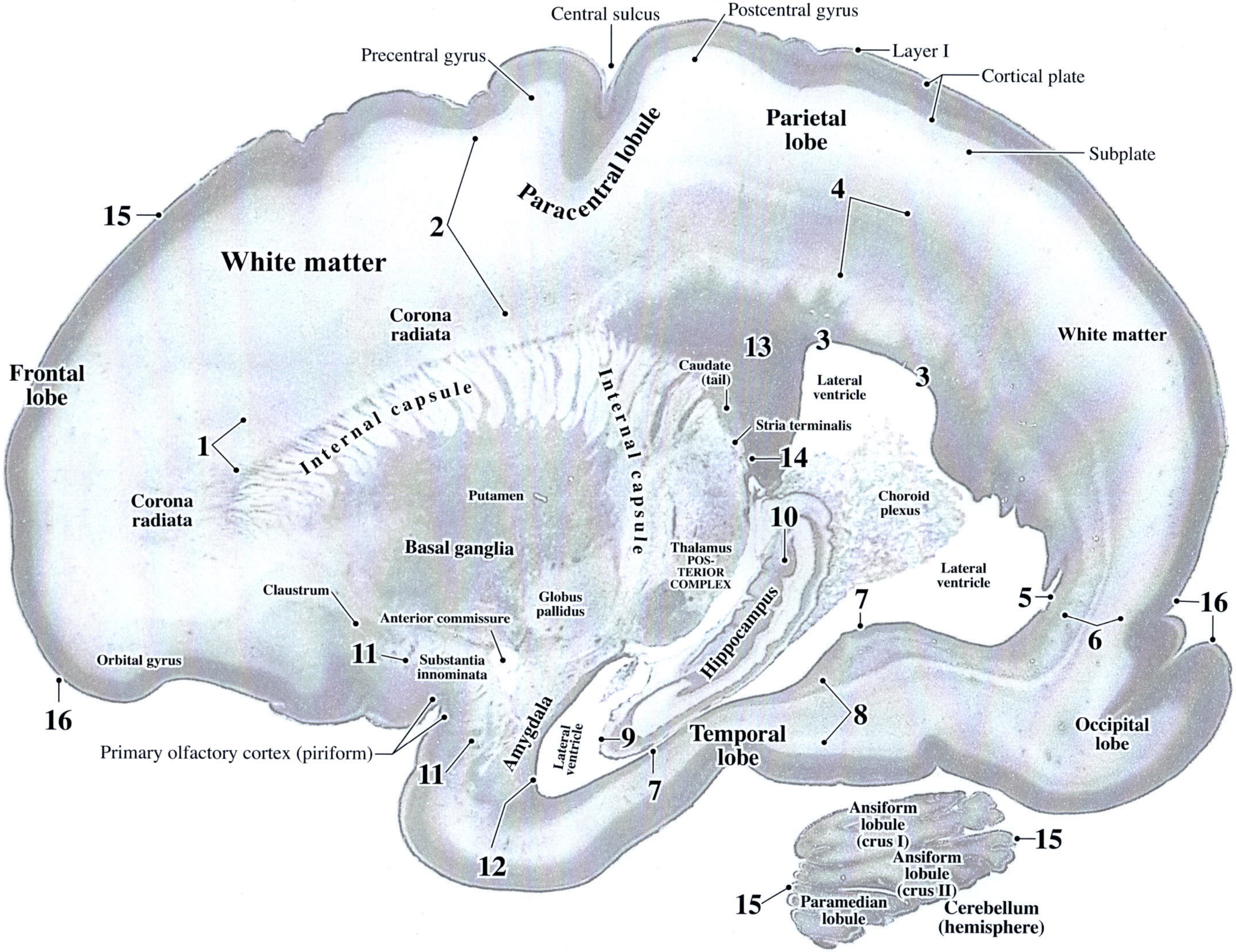
Central sulcus
Postcentral gyrus
Precentral gyrus
Layer I
Cortical plate
Parietal lobe
Paracentral lobule
Subplate
15
4
2
White matter
Corona radiata
13
3
3
White matter
Frontal lobe
Caudate (tail)
Lateral ventricle
Internal capsule
Internal capsule
Stria terminalis
1
14
Corona radiata
Putamen
Choroid plexus
10
Basal ganglia
Thalamus POSTERIOR COMPLEX
Lateral ventricle
Claustrum
Globus pallidus
Hippocampus
7
5
6
16
Anterior commissure
11
Substantia innominata
Orbital gyrus
8
16
Amygdala
Lateral ventricle
9
Temporal lobe
Occipital lobe
Primary olfactory cortex (piriform)
11
7
12
Ansiform lobule (crus I)
15
Ansiform lobule (crus II)
15
Paramedian lobule
Cerebellum (hemisphere)

PLATE 10A
CR 225 mm
GW 26, Y147-63
Sagittal
Section 261

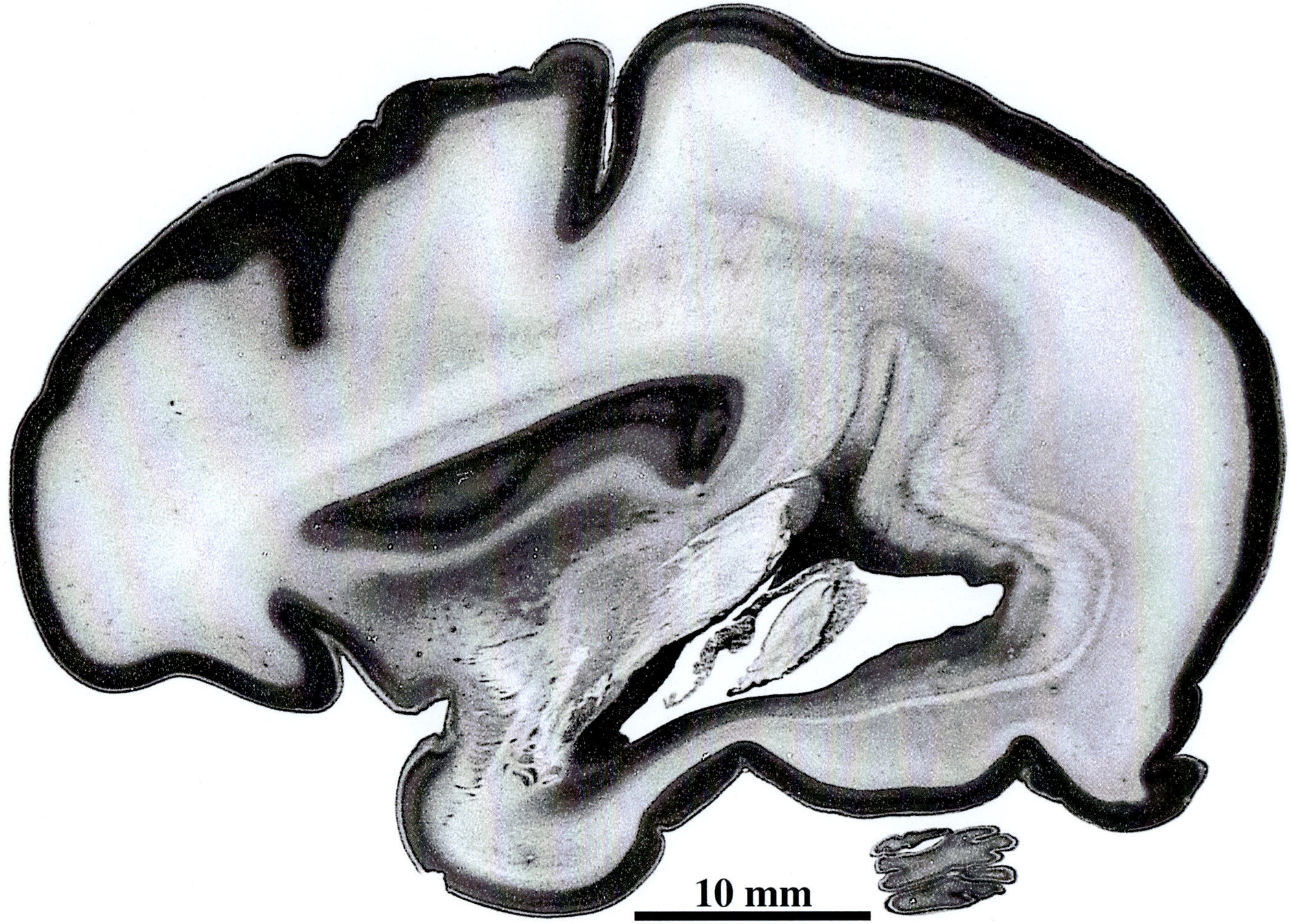

Remnants of the germinal matrix, migratory streams, and transitional fields

1 *Parietal NEP and SVZ*

2 *Parietal STF*

3 *Temporal NEP and SVZ*

4 *Temporal STF*

5 *Lateral migratory stream (cortical)*

6 *Amygdaloid G/EP*

7 *Posterior striatal NEP and SVZ*

8 *Strionuclear GEP*

9 *External germinal layer (cerebellum)*

10 *Subpial granular layer (cortical)*

GEP - Glioepithelium
G/EP - Glioepithelium/ependyma
NEP - Neuroepithelium
STF - Stratified transitional field
SVZ - Subventricular zone

See detail of the brain core in Plates 20A and B.

PLATE 10B

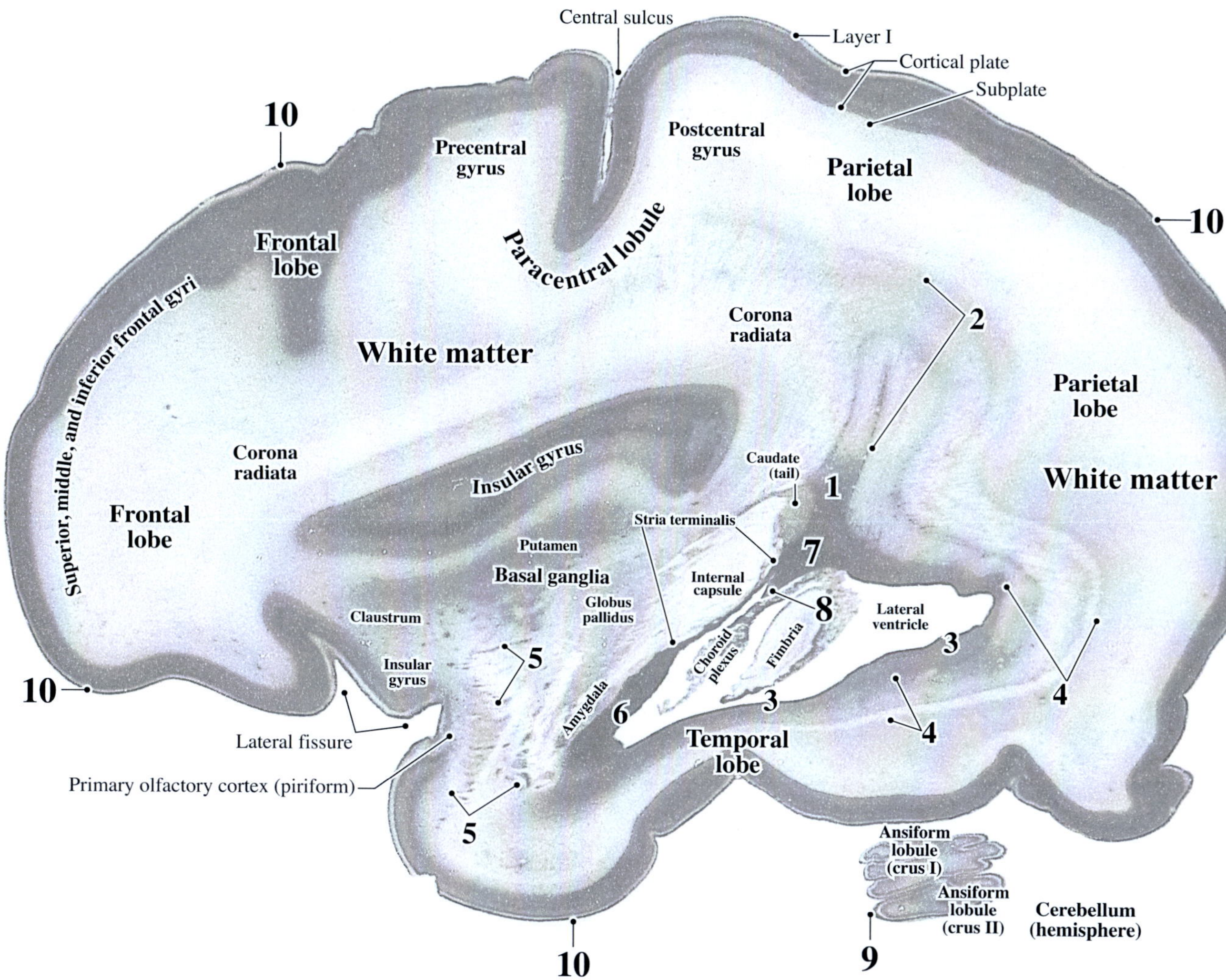
Central sulcus
Layer I
Cortical plate
Subplate
10
Precentral gyrus
Postcentral gyrus
Parietal lobe
10
Frontal lobe
Paracentral lobule
Superior, middle, and inferior frontal gyri
White matter
Corona radiata
2
Parietal lobe
Corona radiata
Insular gyrus
Caudate (tail)
1
White matter
Frontal lobe
Stria terminalis
7
Putamen
Basal ganglia
Internal capsule
8
Globus pallidus
Lateral ventricle
Claustrum
Choroid plexus
Fimbria
3
5
Insular gyrus
4
10
Amygdala
6
3
4
Lateral fissure
Temporal lobe
Primary olfactory cortex (piriform)
5
Ansiform lobule (crus I)
Ansiform lobule (crus II)
Cerebellum (hemisphere)
10
9

PLATE 11A
CR 225 mm
GW 26, Y147-63
Sagittal
Section 501

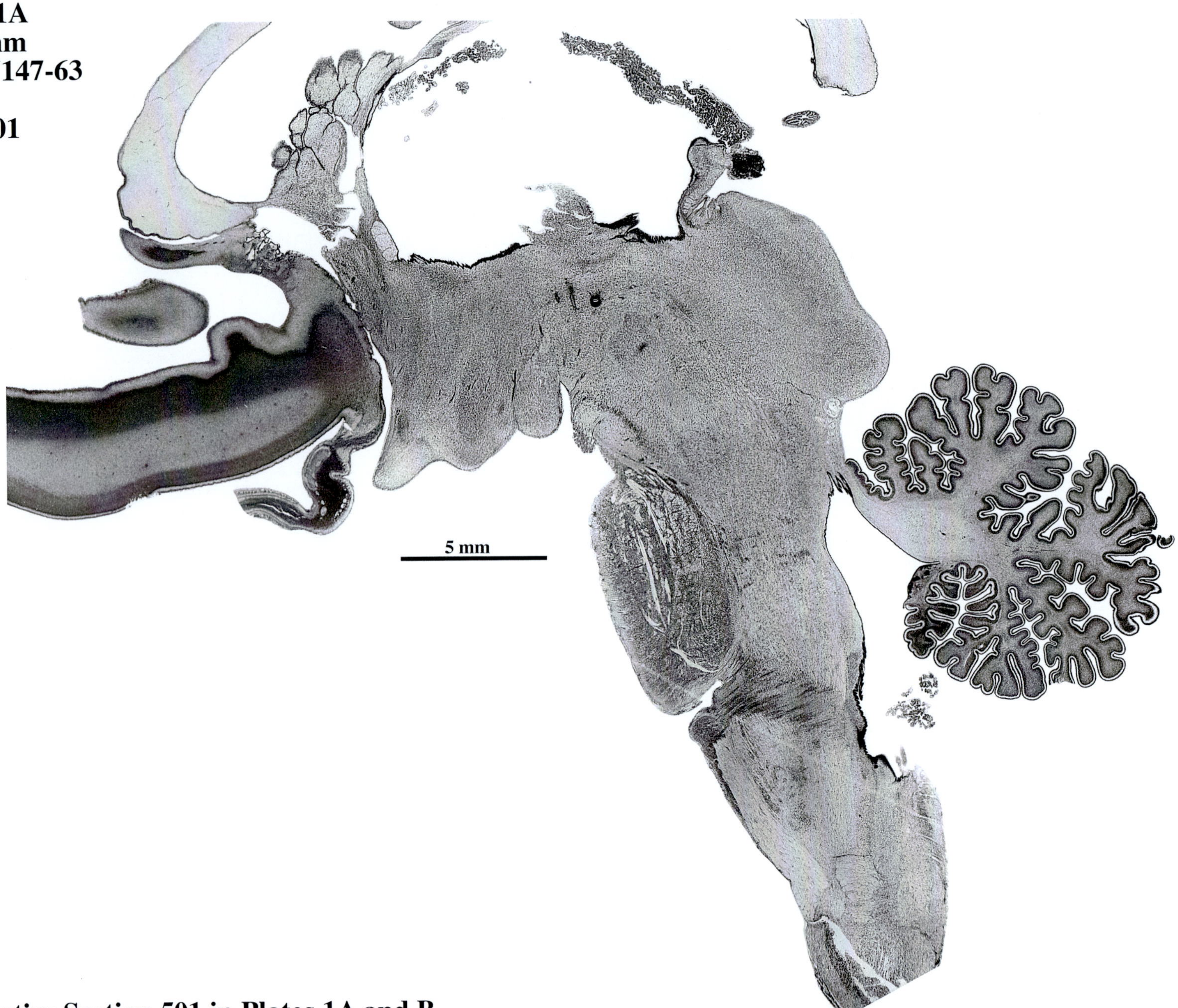

See the entire Section 501 in Plates 1A and B.

PLATE 11B

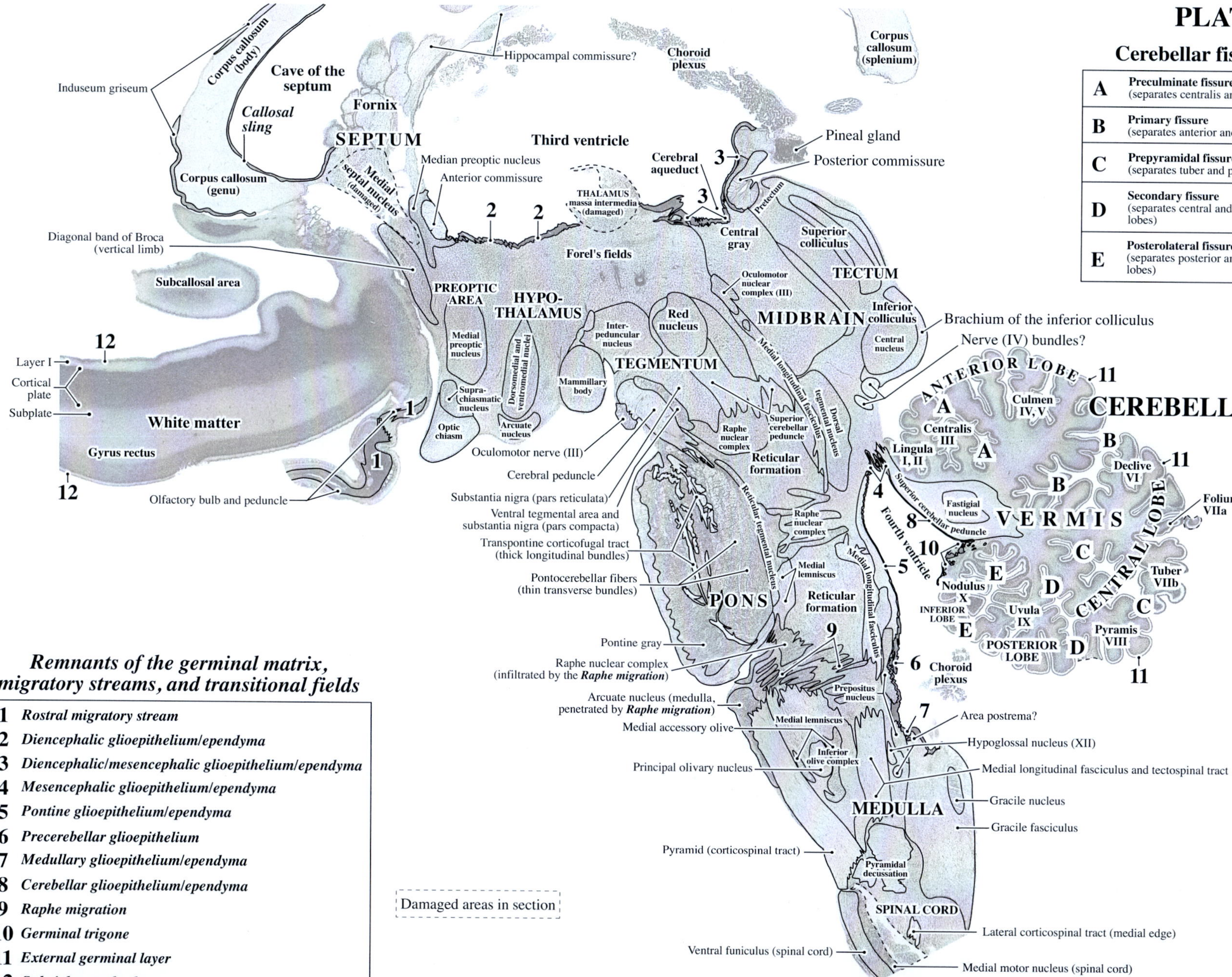

Cerebellar fissures

A	**Preculminate fissure** (separates centralis and culmen)
B	**Primary fissure** (separates anterior and central lobes)
C	**Prepyramidal fissure** (separates tuber and pyramis)
D	**Secondary fissure** (separates central and posterior lobes)
E	**Posterolateral fissure** (separates posterior and inferior lobes)

Remnants of the germinal matrix, migratory streams, and transitional fields

1	***Rostral migratory stream***
2	***Diencephalic glioepithelium/ependyma***
3	***Diencephalic/mesencephalic glioepithelium/ependyma***
4	***Mesencephalic glioepithelium/ependyma***
5	***Pontine glioepithelium/ependyma***
6	***Precerebellar glioepithelium***
7	***Medullary glioepithelium/ependyma***
8	***Cerebellar glioepithelium/ependyma***
9	***Raphe migration***
10	***Germinal trigone***
11	***External germinal layer***
12	***Subpial granular layer***

PLATE 12A
CR 225 mm
GW 26, Y147-63
Sagittal
Section 481

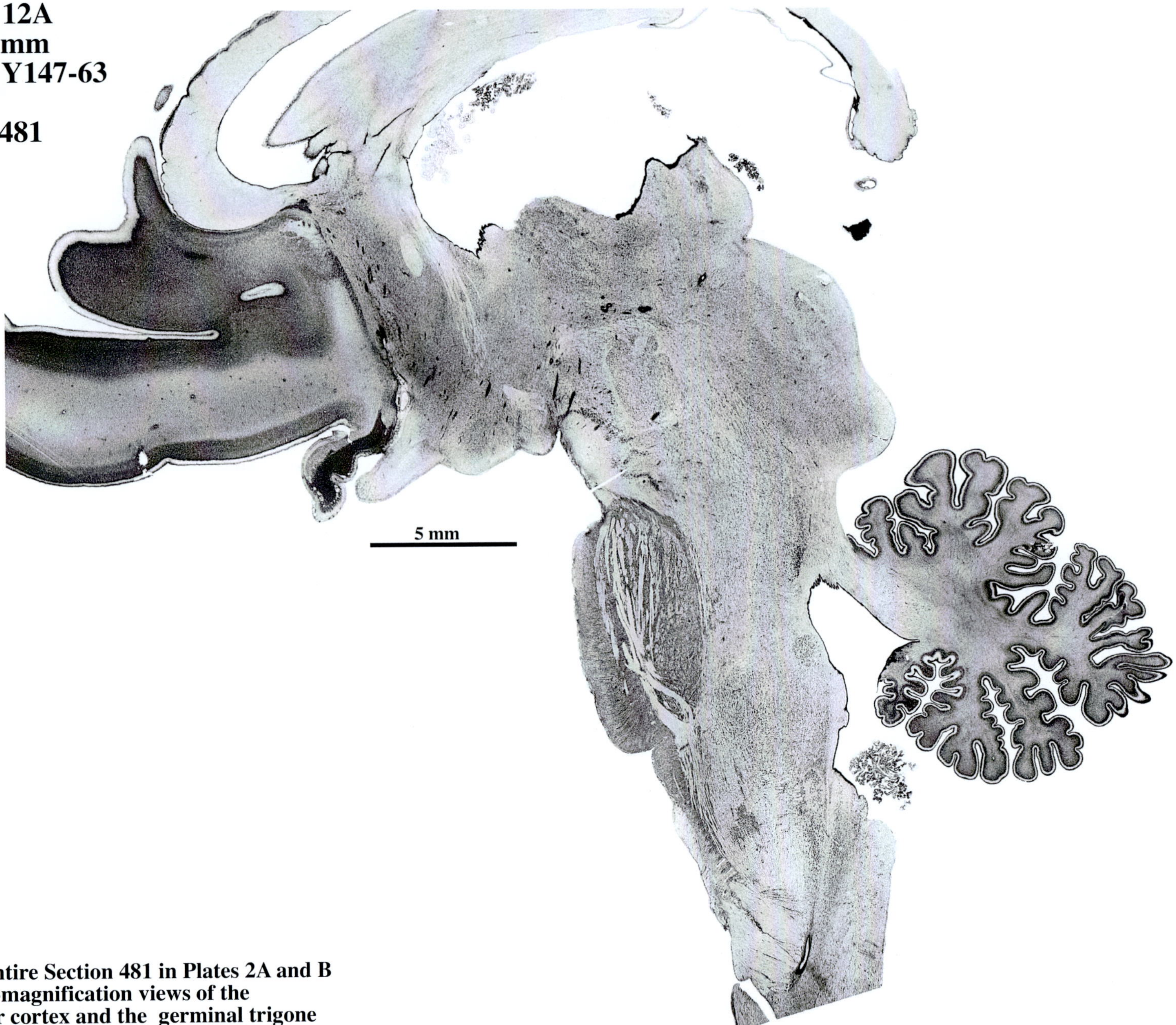

See the entire Section 481 in Plates 2A and B and high-magnification views of the cerebellar cortex and the germinal trigone in Plates 21-26.

PLATE 12B

Cerebellar fissures

A	**Preculminate fissure** (separates centralis and culmen)
B	**Primary fissure** (separates anterior and central lobes)
C	**Prepyramidal fissure** (separates tuber and pyramis)
D	**Secondary fissure** (separates central and posterior lobes)
E	**Posterolateral fissure** (separates posterior and inferior lobes)

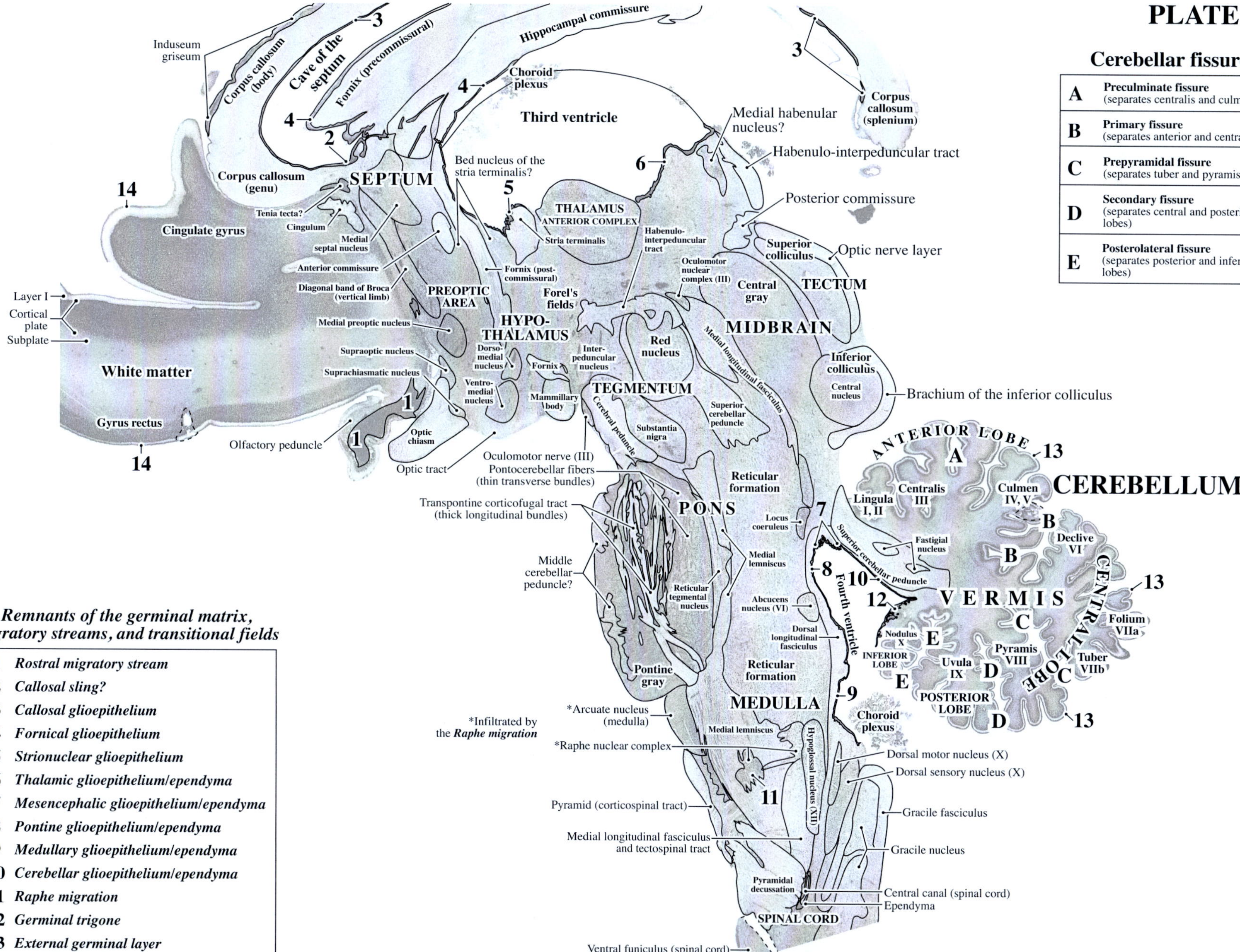

Remnants of the germinal matrix, migratory streams, and transitional fields

1	*Rostral migratory stream*
2	*Callosal sling?*
3	*Callosal glioepithelium*
4	*Fornical glioepithelium*
5	*Strionuclear glioepithelium*
6	*Thalamic glioepithelium/ependyma*
7	*Mesencephalic glioepithelium/ependyma*
8	*Pontine glioepithelium/ependyma*
9	*Medullary glioepithelium/ependyma*
10	*Cerebellar glioepithelium/ependyma*
11	*Raphe migration*
12	*Germinal trigone*
13	*External germinal layer*
14	*Subpial granular layer*

Damaged areas in section

PLATE 13A
CR 225 mm
GW 26
Y147-63
Sagittal
Section
461

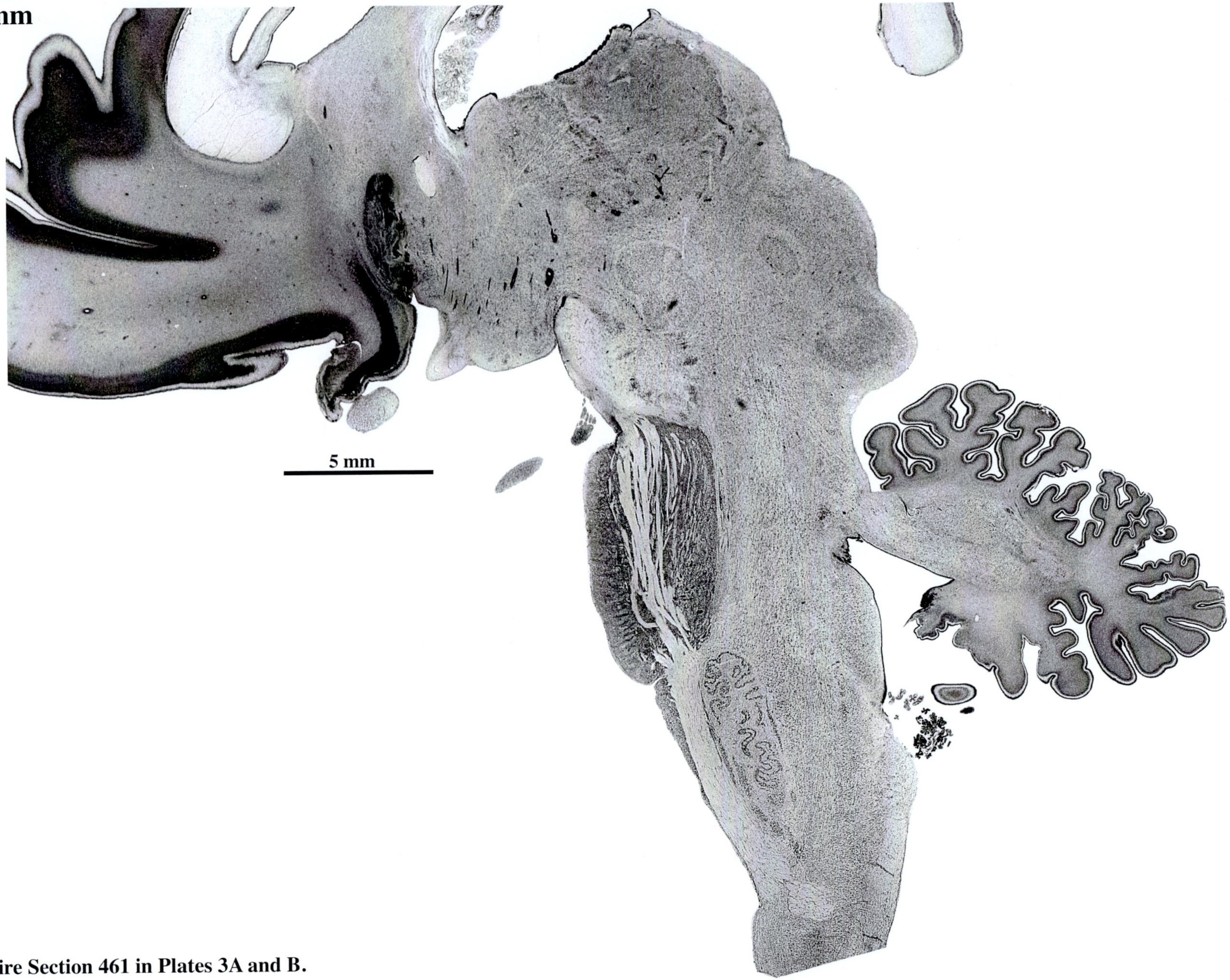

See the entire Section 461 in Plates 3A and B.

PLATE 13B

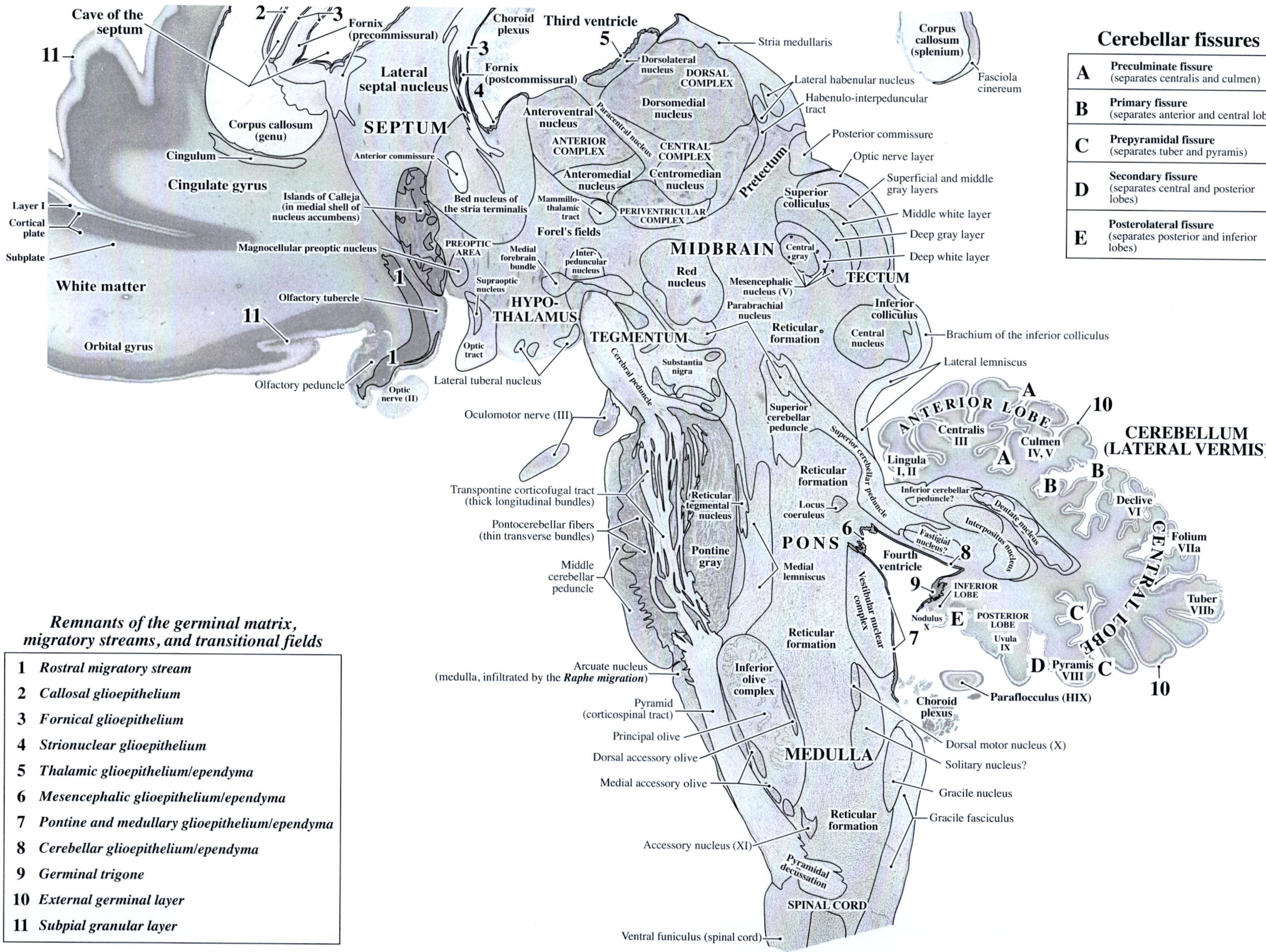

Cerebellar fissures

A	**Preculminate fissure** (separates centralis and culmen)
B	**Primary fissure** (separates anterior and central lobes)
C	**Prepyramidal fissure** (separates tuber and pyramis)
D	**Secondary fissure** (separates central and posterior lobes)
E	**Posterolateral fissure** (separates posterior and inferior lobes)

Remnants of the germinal matrix, migratory streams, and transitional fields

1	*Rostral migratory stream*
2	*Callosal glioepithelium*
3	*Fornical glioepithelium*
4	*Strionuclear glioepithelium*
5	*Thalamic glioepithelium/ependyma*
6	*Mesencephalic glioepithelium/ependyma*
7	*Pontine and medullary glioepithelium/ependyma*
8	*Cerebellar glioepithelium/ependyma*
9	*Germinal trigone*
10	*External germinal layer*
11	*Subpial granular layer*

PLATE 14A
CR 225 mm
GW 26
Y147-63
Sagittal
Section
421

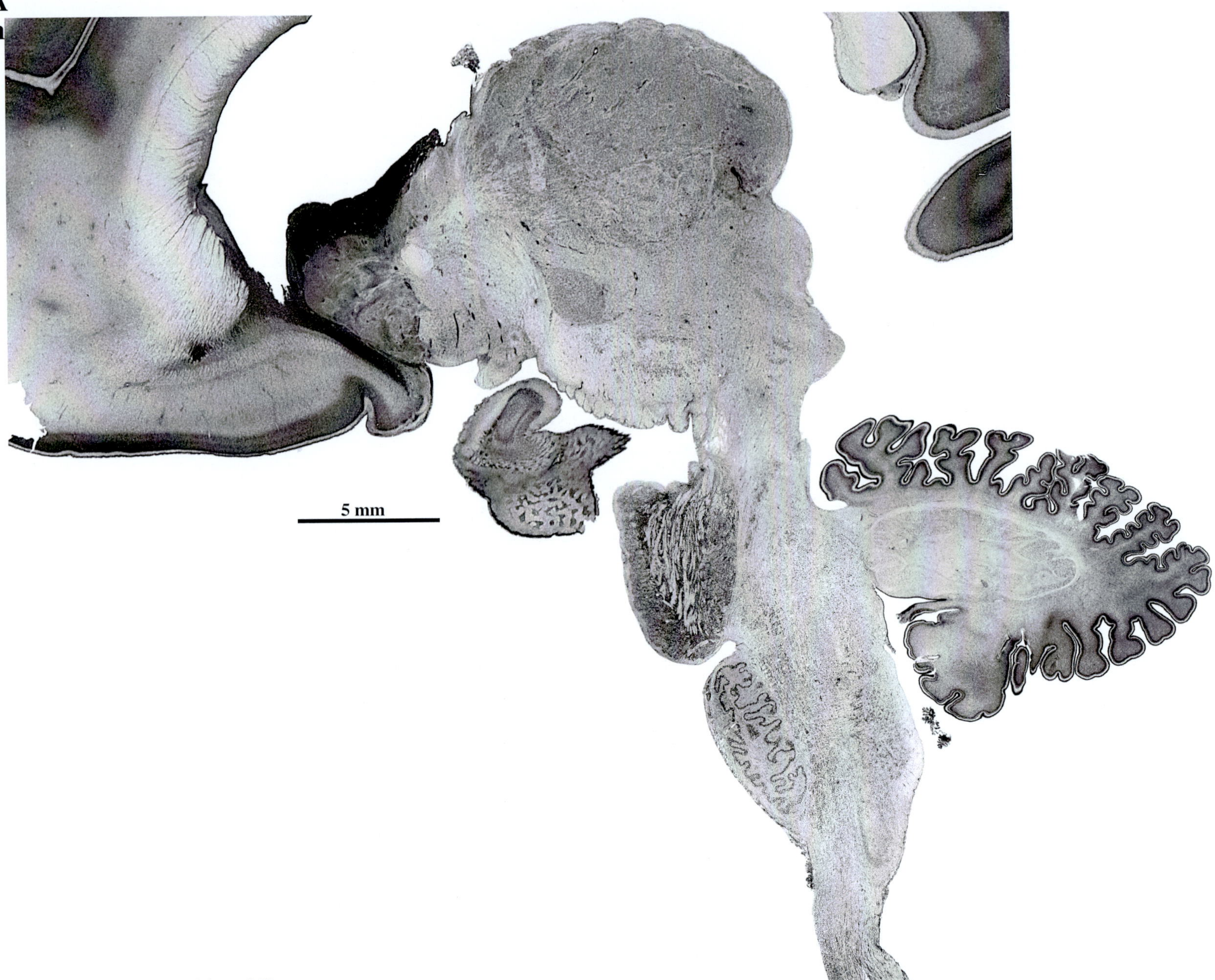

See the entire Section 421 in Plates 4A and B.

PLATE 14B

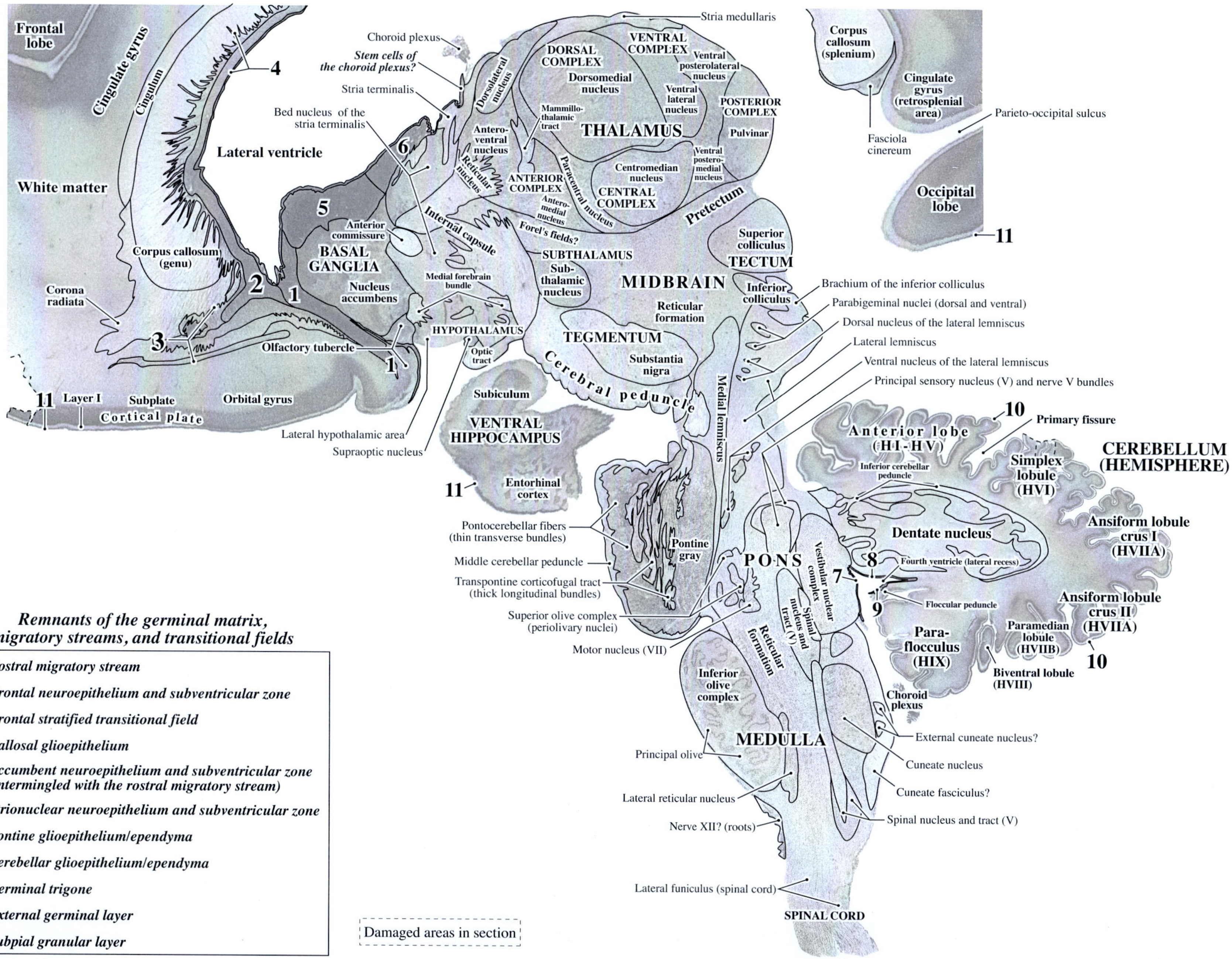

Remnants of the germinal matrix, migratory streams, and transitional fields

1	*Rostral migratory stream*
2	*Frontal neuroepithelium and subventricular zone*
3	*Frontal stratified transitional field*
4	*Callosal glioepithelium*
5	*Accumbent neuroepithelium and subventricular zone (intermingled with the rostral migratory stream)*
6	*Strionuclear neuroepithelium and subventricular zone*
7	*Pontine glioepithelium/ependyma*
8	*Cerebellar glioepithelium/ependyma*
9	*Germinal trigone*
10	*External germinal layer*
11	*Subpial granular layer*

PLATE 15A, CR 225 mm, GW 26, Y147-63, Sagittal, Section 381

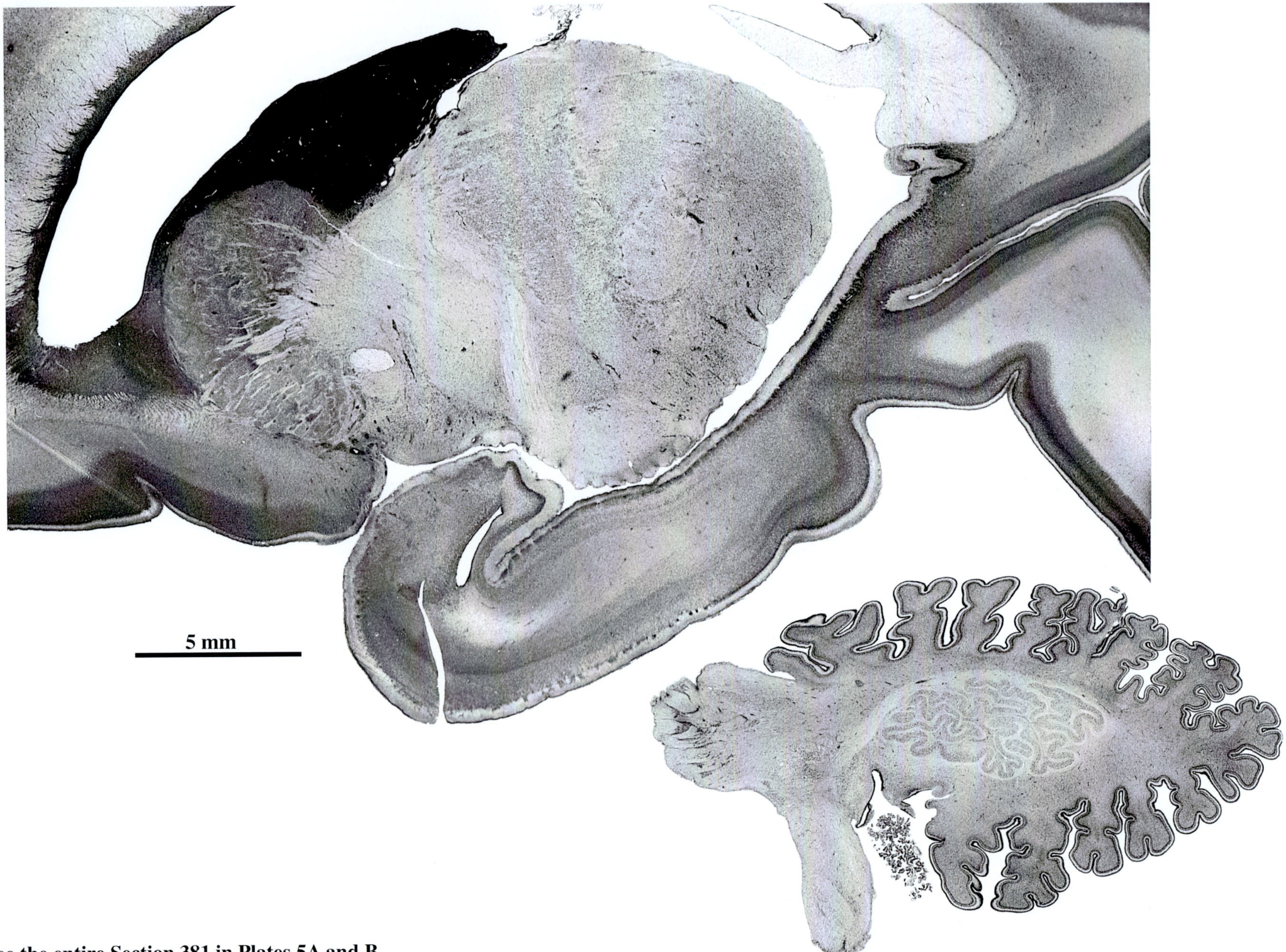

See the entire Section 381 in Plates 5A and B.

PLATE 15B

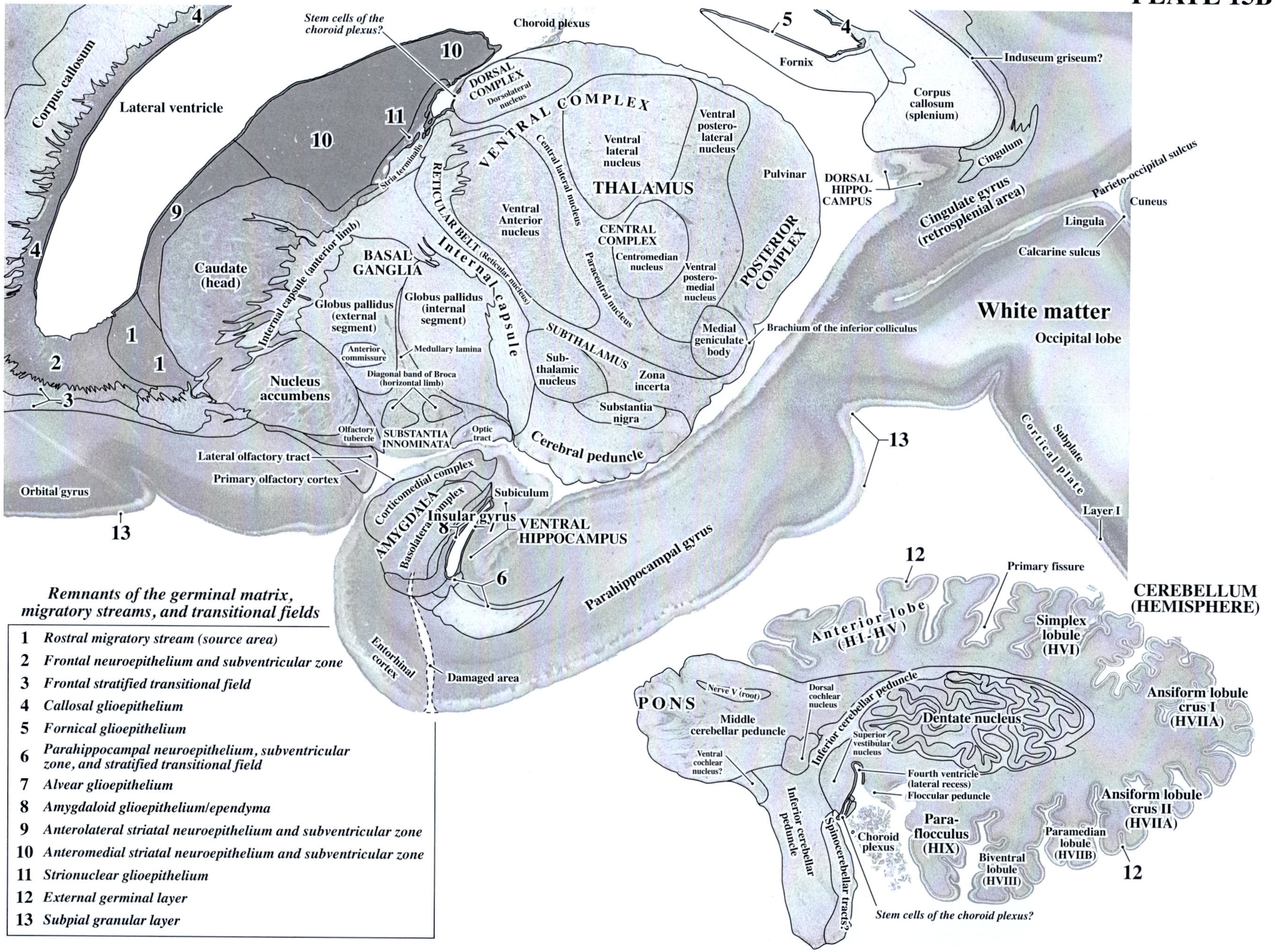

Remnants of the germinal matrix, migratory streams, and transitional fields

1 *Rostral migratory stream (source area)*
2 *Frontal neuroepithelium and subventricular zone*
3 *Frontal stratified transitional field*
4 *Callosal glioepithelium*
5 *Fornical glioepithelium*
6 *Parahippocampal neuroepithelium, subventricular zone, and stratified transitional field*
7 *Alvear glioepithelium*
8 *Amygdaloid glioepithelium/ependyma*
9 *Anterolateral striatal neuroepithelium and subventricular zone*
10 *Anteromedial striatal neuroepithelium and subventricular zone*
11 *Strionuclear glioepithelium*
12 *External germinal layer*
13 *Subpial granular layer*

PLATE 16A
CR 225 mm
GW 26
Y147-63
Sagittal
Section 361

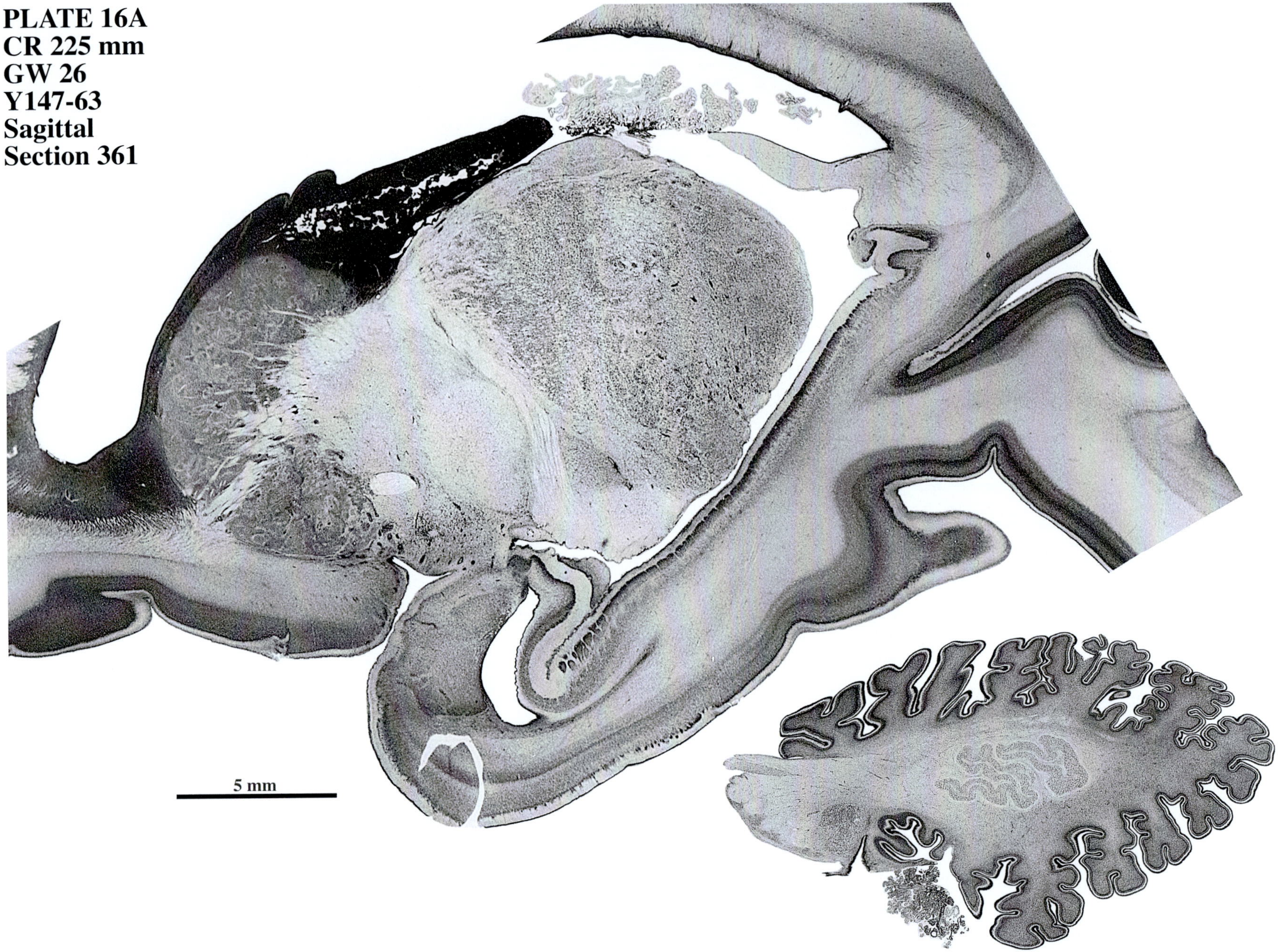

See the entire Section 361 in Plates 6A and B.

PLATE 16B

Remnants of the germinal matrix, migratory streams, and transitional fields

1. *Rostral migratory stream (source area)*
2. *Frontal neuroepithelium and subventricular zone*
3. *Frontal stratified transitional field*
4. *Callosal glioepithelium*
5. *Fornical glioepithelium*
6. *Occipital neuroepithelium and subventricular zone*
7. *Occipital stratified transitional field*

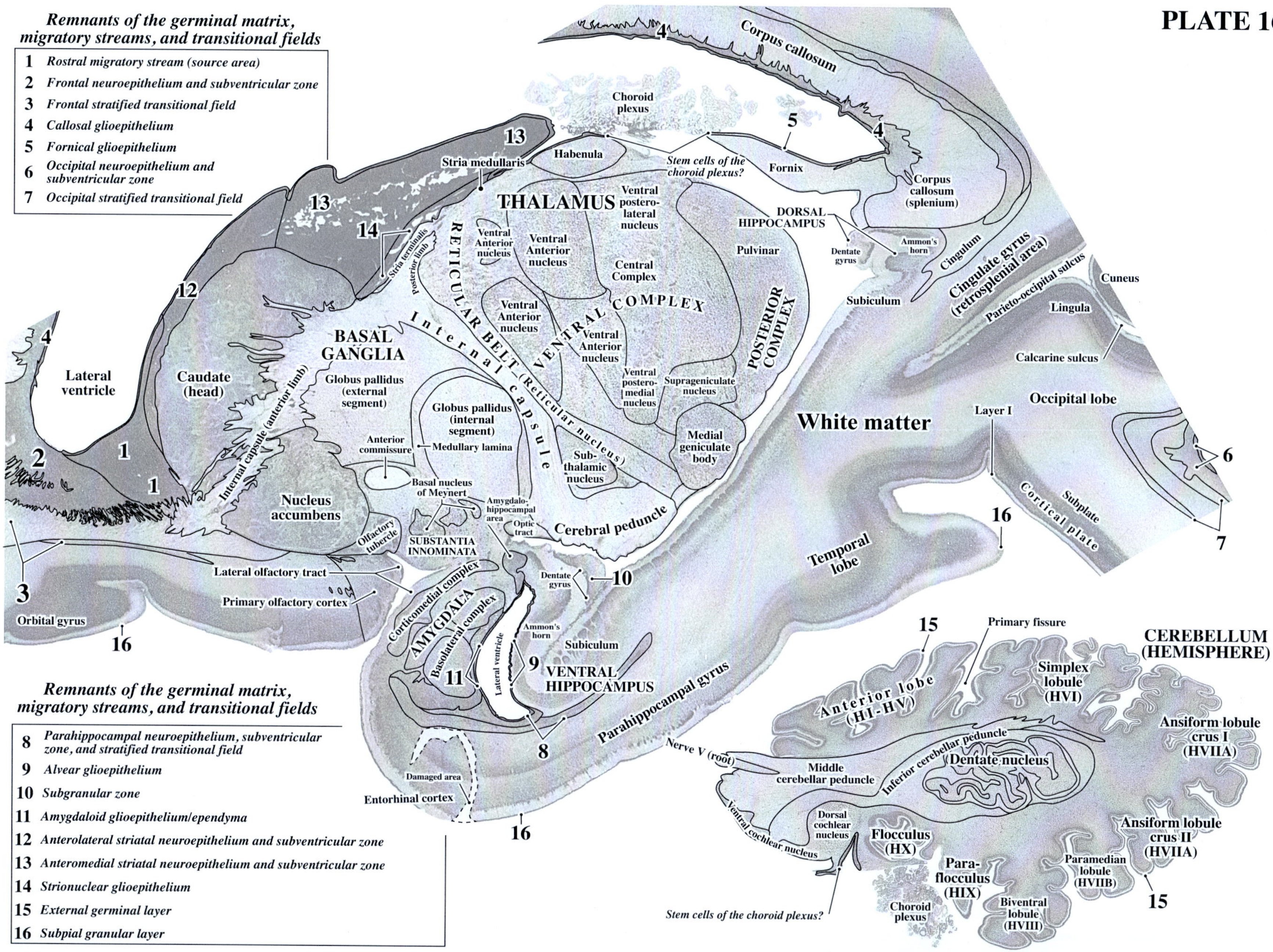

Remnants of the germinal matrix, migratory streams, and transitional fields

8. *Parahippocampal neuroepithelium, subventricular zone, and stratified transitional field*
9. *Alvear glioepithelium*
10. *Subgranular zone*
11. *Amygdaloid glioepithelium/ependyma*
12. *Anterolateral striatal neuroepithelium and subventricular zone*
13. *Anteromedial striatal neuroepithelium and subventricular zone*
14. *Strionuclear glioepithelium*
15. *External germinal layer*
16. *Subpial granular layer*

PLATE 17A
CR 225 mm
GW 26
Y147-63
Sagittal
Section 341

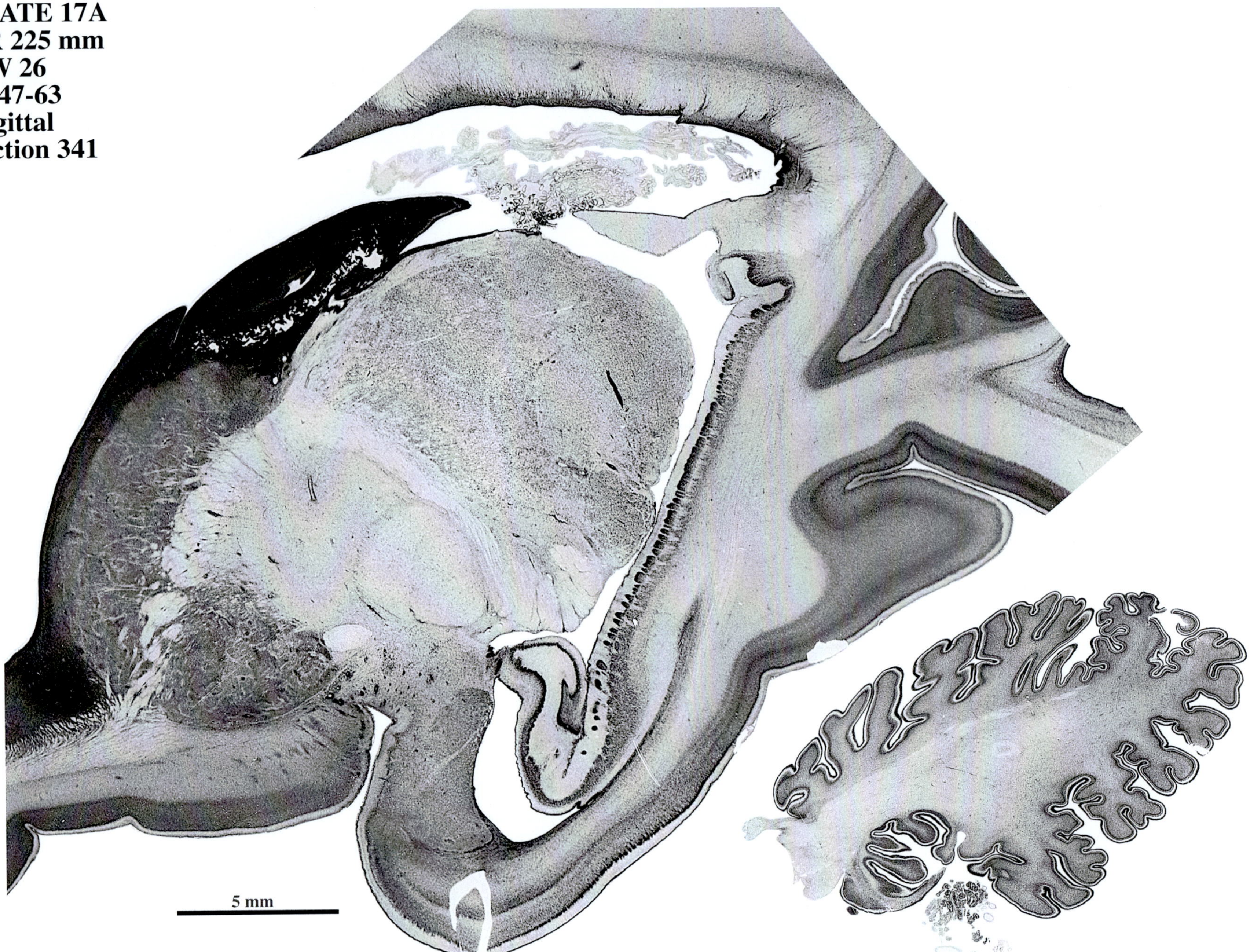

See the entire Section 341 in Plates 7A and B.

PLATE 17B

Germinal and transitional structures in *italics*

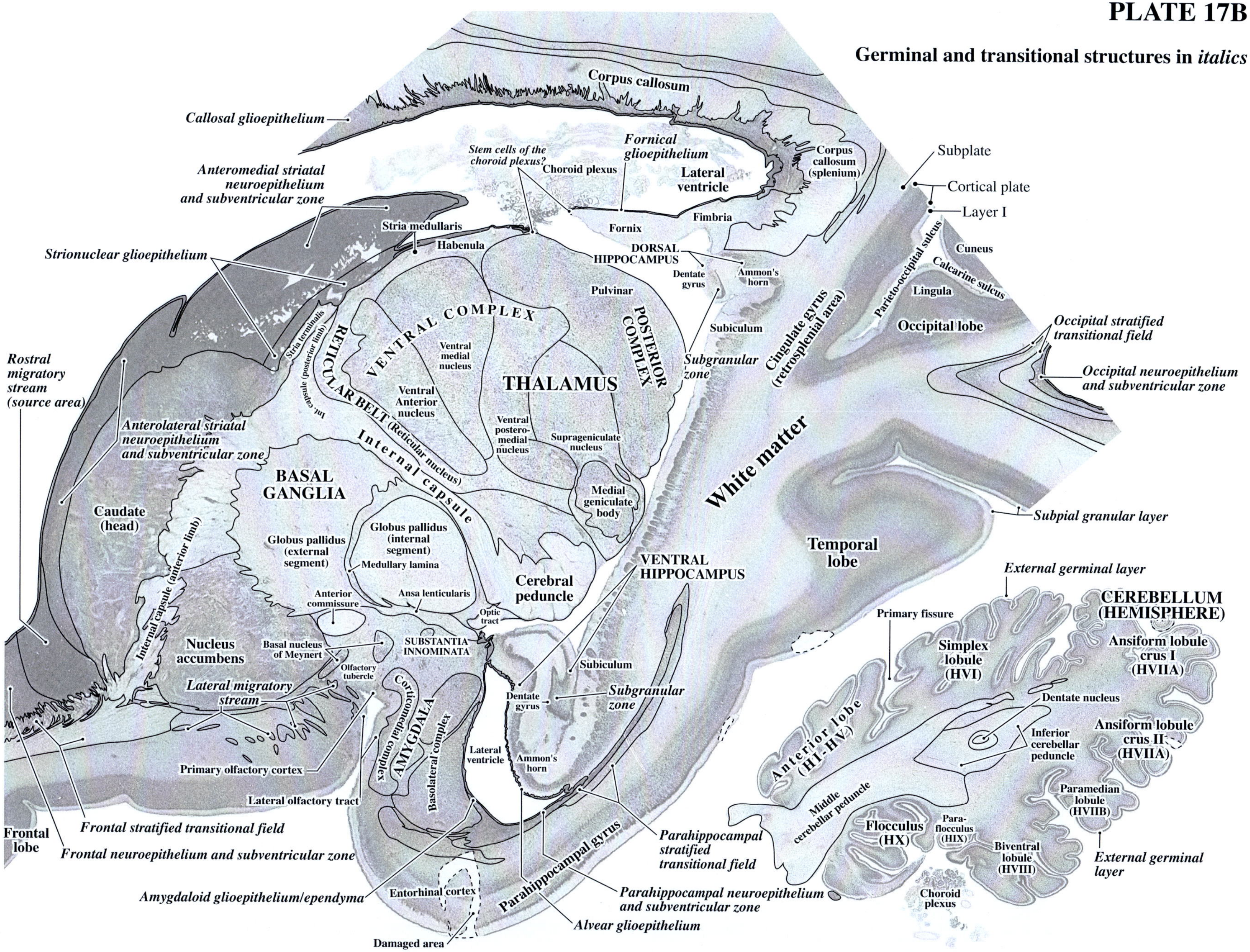

PLATE 18A, CR 225 mm, GW 26, Y147-63, Sagittal, Section 341

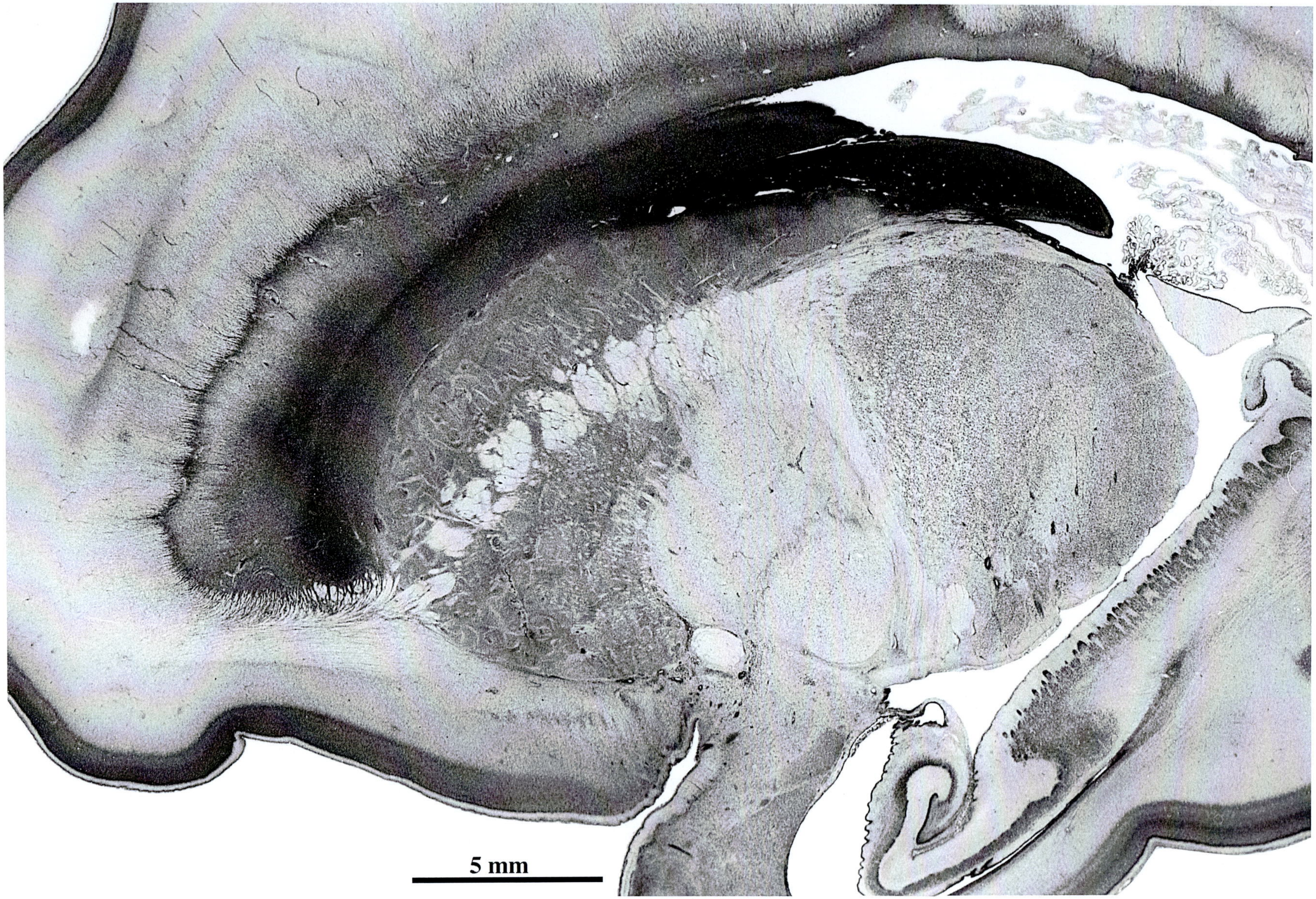

See the entire Section 321 in Plates 8A and B.

PLATE 18B

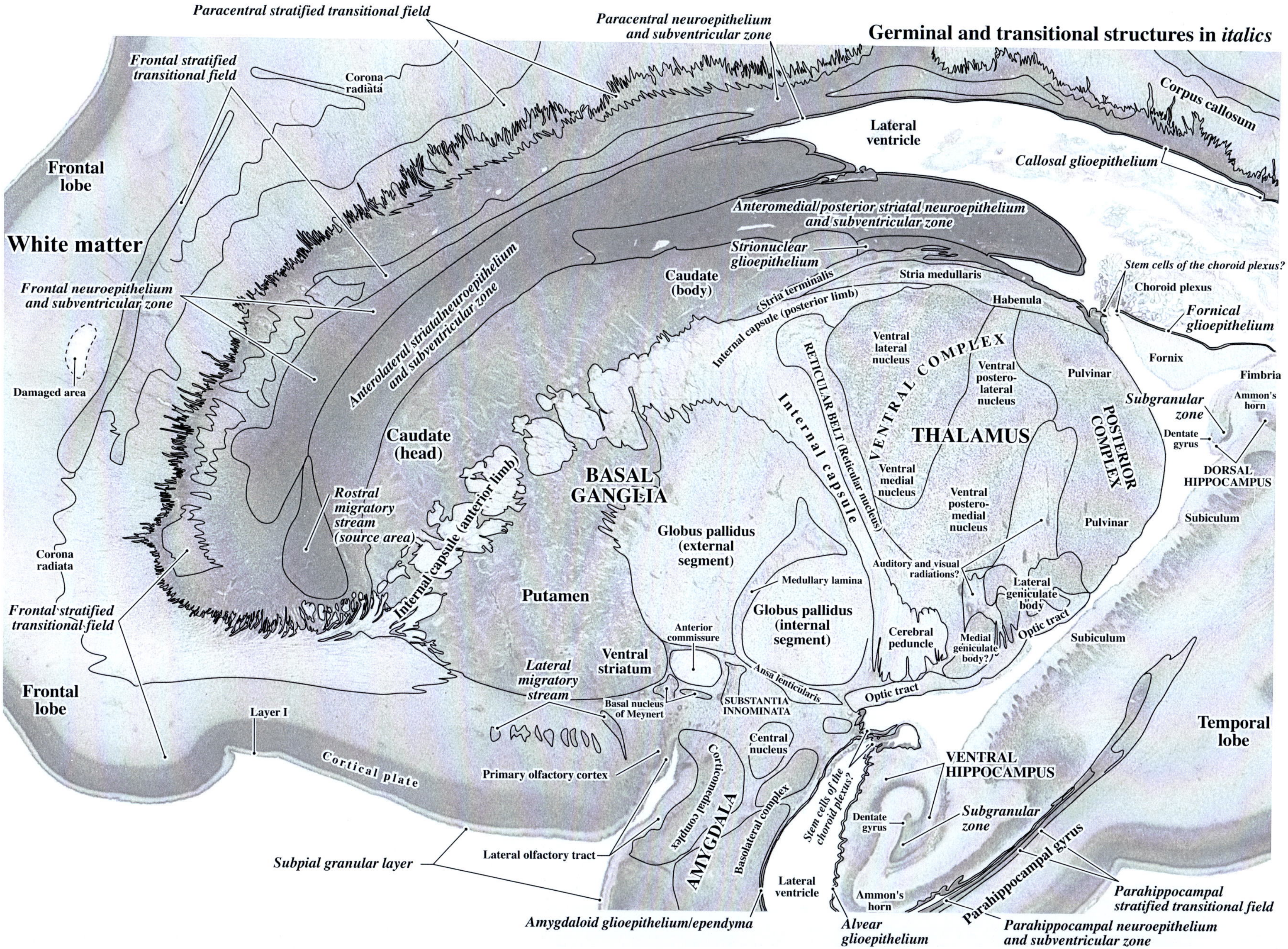

Germinal and transitional structures in *italics*
Paracentral stratified transitional field
Paracentral neuroepithelium and subventricular zone
Frontal stratified transitional field
Corona radiata
Lateral ventricle
Corpus callosum
Callosal glioepithelium
Frontal lobe
White matter
Anteromedial/posterior striatal neuroepithelium and subventricular zone
Strionuclear glioepithelium
Stem cells of the choroid plexus?
Choroid plexus
Fornical glioepithelium
Stria medullaris
Frontal neuroepithelium and subventricular zone
Caudate (body)
Stria terminalis
Habenula
Internal capsule (posterior limb)
Damaged area
Anterolateral striatalneuroepithelium and subventricular zone
Ventral lateral nucleus
VENTRAL COMPLEX
Ventral postero-lateral nucleus
Fornix
Pulvinar
Fimbria
RETICULAR BELT (Reticular nucleus)
Internal capsule
Subgranular zone
Ammon's horn
Caudate (head)
THALAMUS
POSTERIOR COMPLEX
Dentate gyrus
BASAL GANGLIA
Ventral medial nucleus
DORSAL HIPPOCAMPUS
Rostral migratory stream (source area)
Internal capsule (anterior limb)
Ventral postero-medial nucleus
Subiculum
Corona radiata
Globus pallidus (external segment)
Pulvinar
Auditory and visual radiations?
Lateral geniculate body
Medullary lamina
Frontal stratified transitional field
Putamen
Globus pallidus (internal segment)
Optic tract
Anterior commissure
Cerebral peduncle
Medial geniculate body?
Subiculum
Ventral striatum
Lateral migratory stream
Ansa lenticularis
Frontal lobe
Basal nucleus of Meynert
SUBSTANTIA INNOMINATA
Optic tract
Layer I
Temporal lobe
Central nucleus
Cortical plate
Primary olfactory cortex
Corticomedial complex
VENTRAL HIPPOCAMPUS
AMYGDALA
Basolateral complex
Stem cells of the choroid plexus?
Dentate gyrus
Subgranular zone
Parahippocampal gyrus
Lateral olfactory tract
Subpial granular layer
Lateral ventricle
Ammon's horn
Parahippocampal stratified transitional field
Amygdaloid glioepithelium/ependyma
Alvear glioepithelium
Parahippocampal neuroepithelium and subventricular zone

PLATE 19A
CR 225 mm, GW 26, Y147-63, Sagittal, Section 261

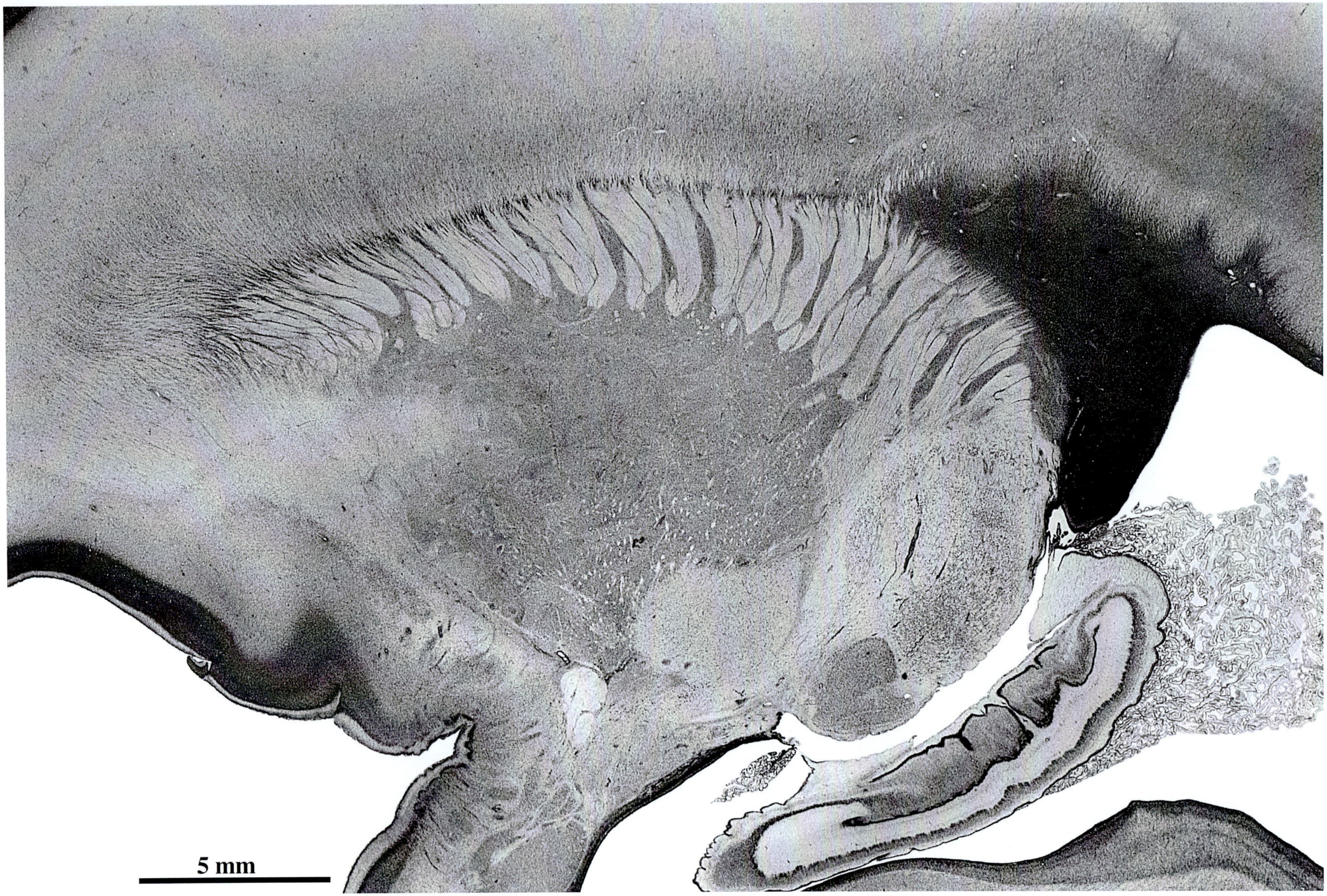

See the entire Section 261 in Plates 9A and B.

PLATE 19B

Germinal and transitional structures in *italics*

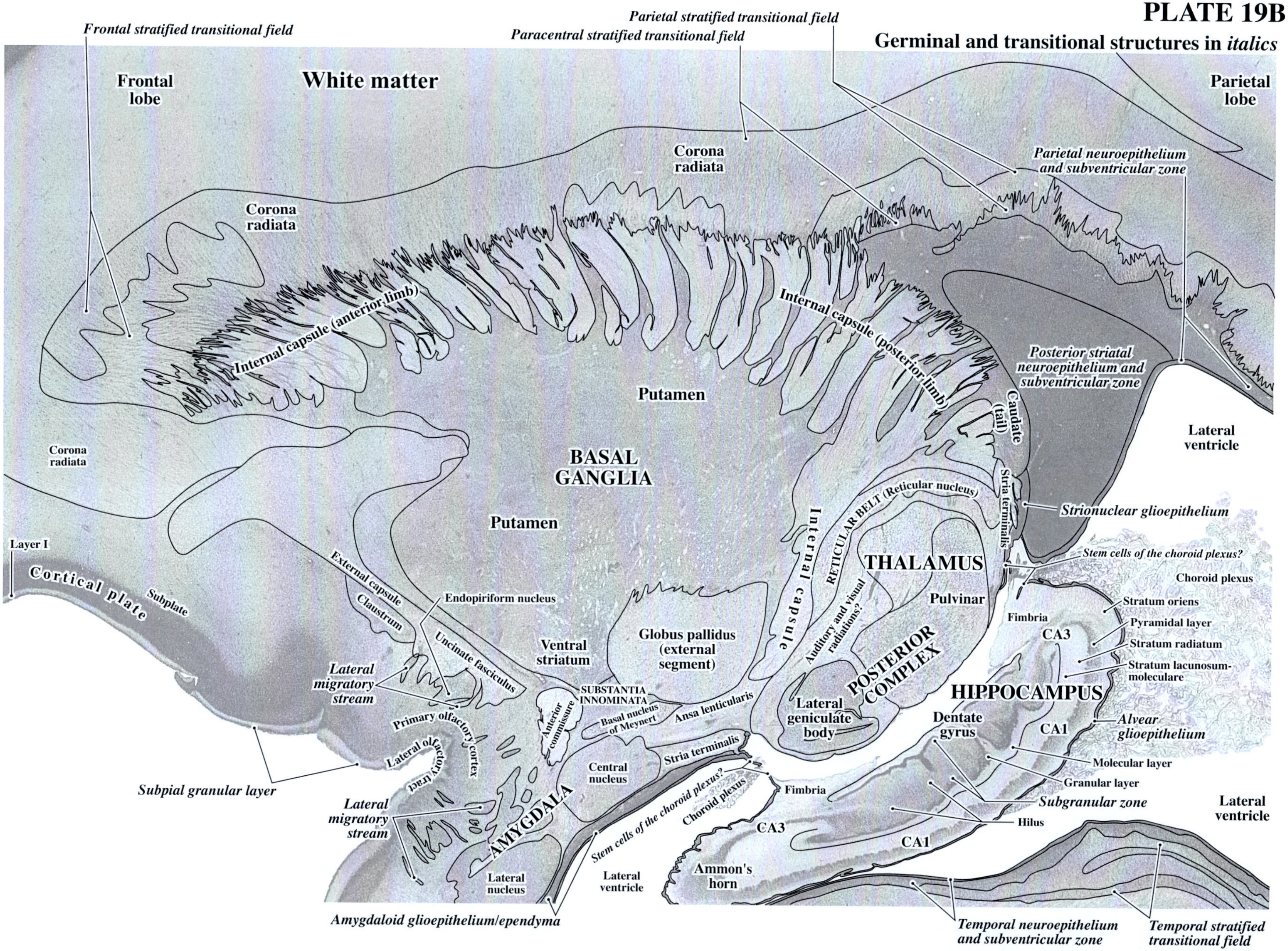

PLATE 20A
CR 225 mm, GW 26, Y147-63, Sagittal, Section 221

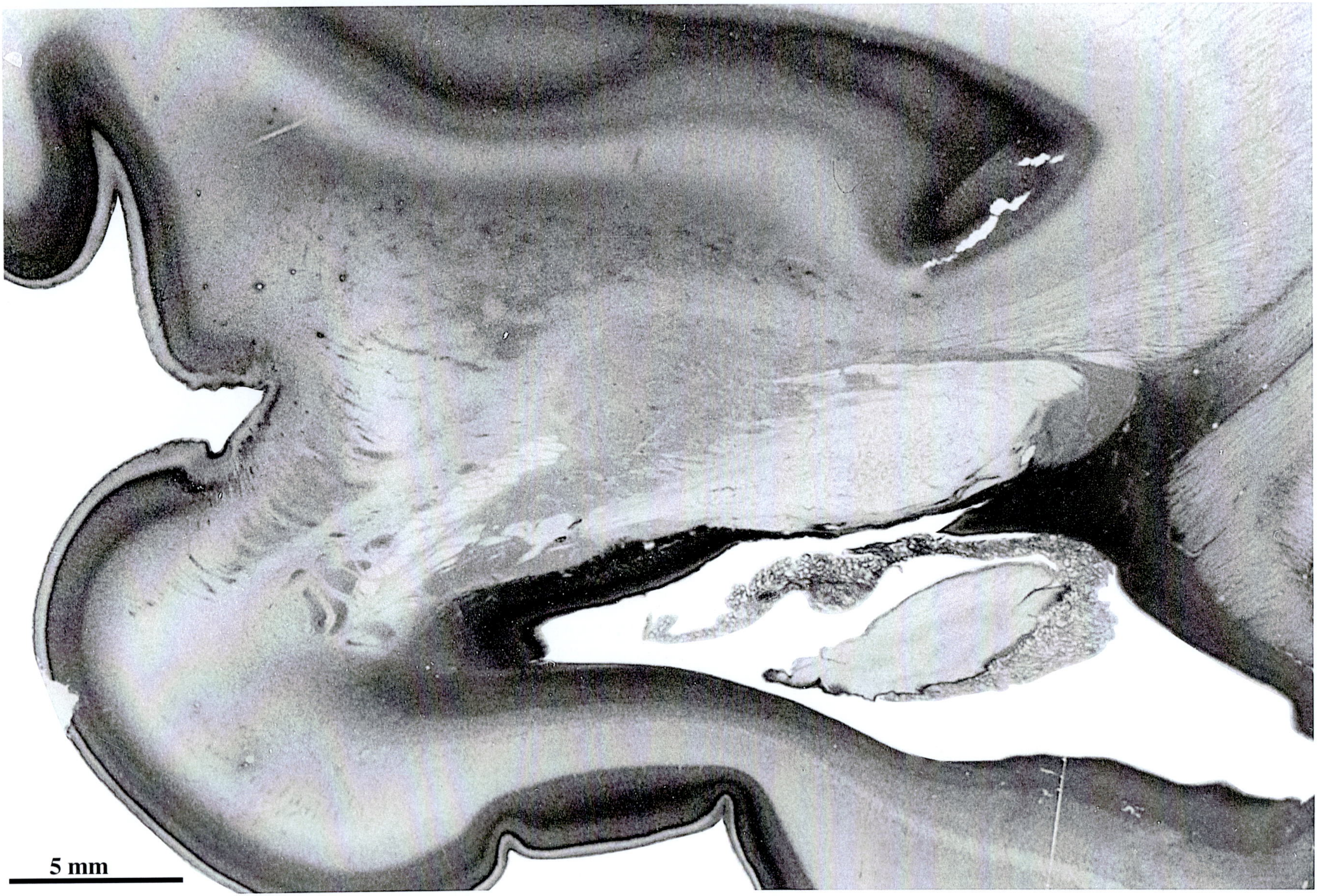

See the entire Section 221 in Plates 10A and B.

PLATE 20B

Germinal and transitional structures in *italics*

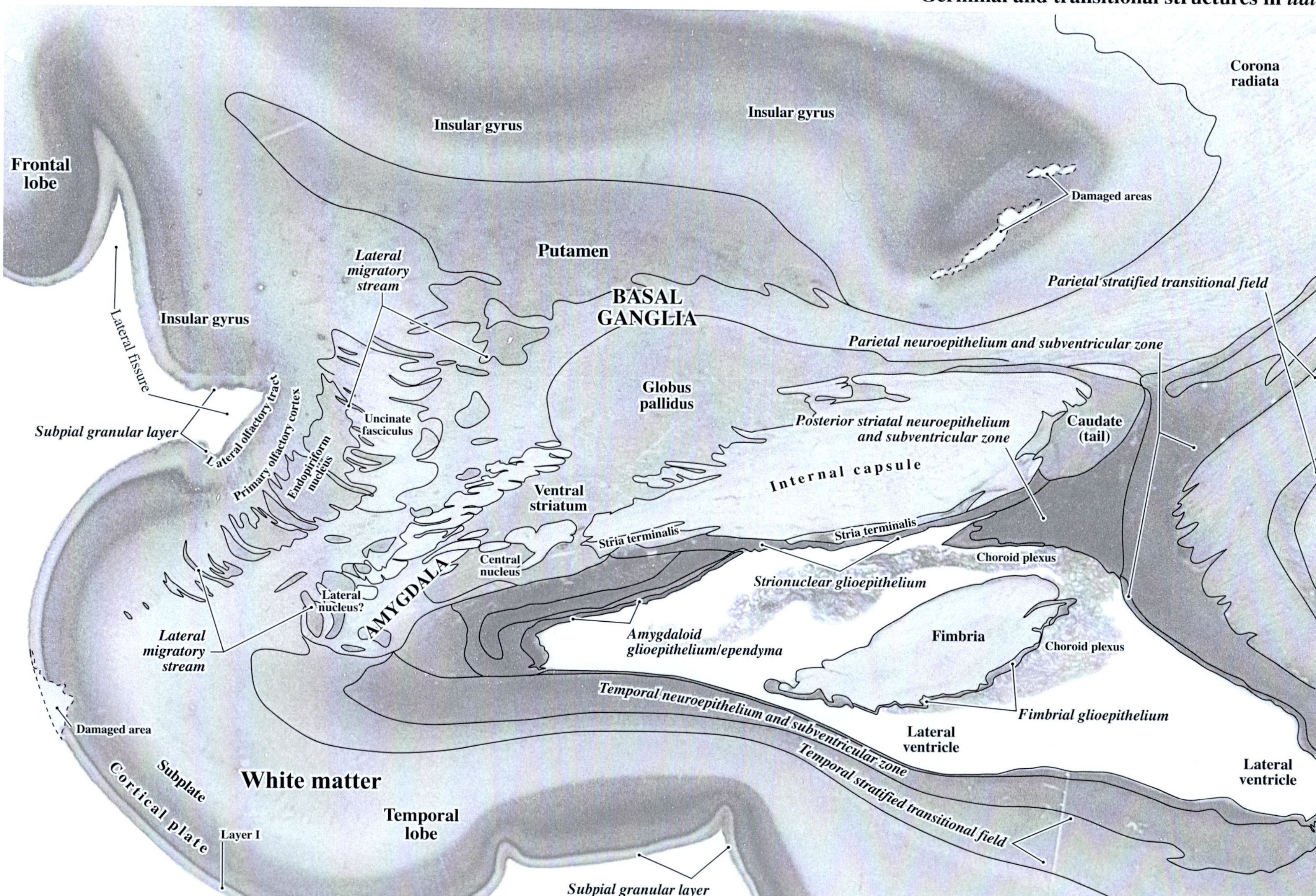

PLATE 21

CR 225 mm, GW 26, Y147-63, Sagittal, Section 481

CEREBELLUM, ANTERIOR LOBE, LINGULA

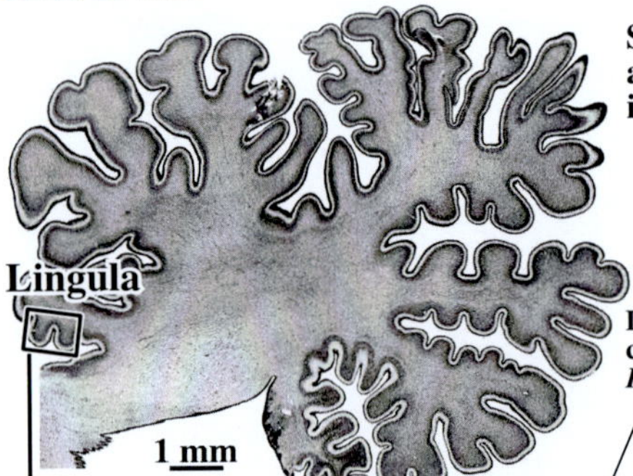

See the entire Section 481 in Plates 2A and B and a medium-magnification view in Plates 12A and B.

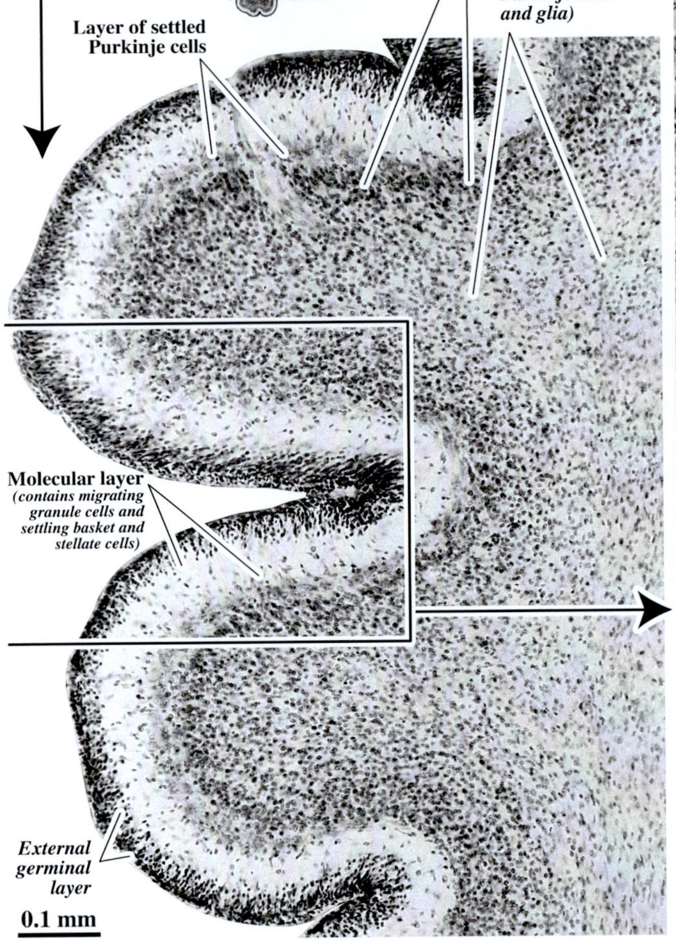

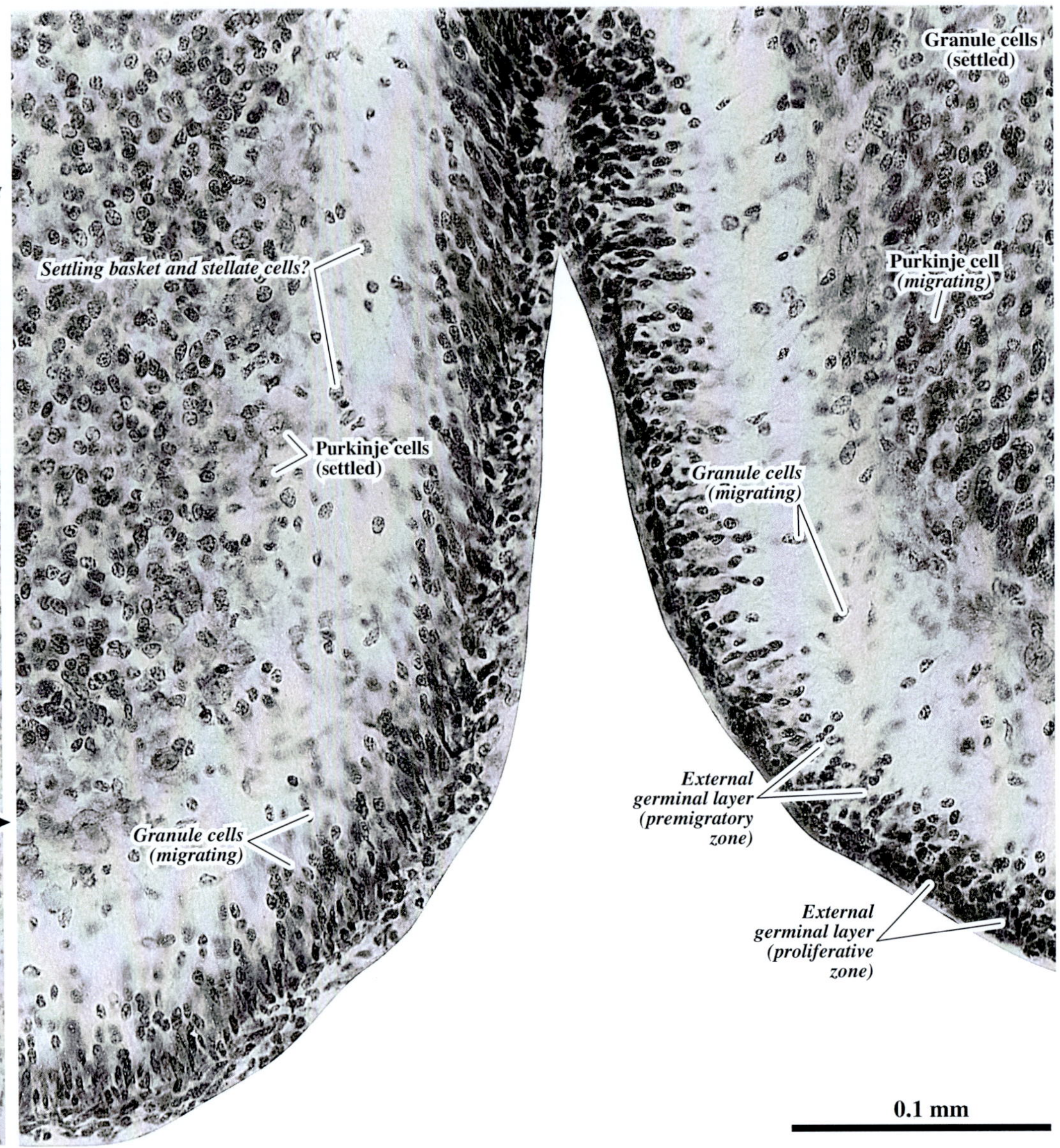

PLATE 22

CR 225 mm, GW 26, Y147-63, Sagittal, Section 481

CEREBELLUM, ANTERIOR LOBE, CULMEN

See the entire Section 481 in Plates 2A and B and a medium-magnification view in Plates 12A and B.

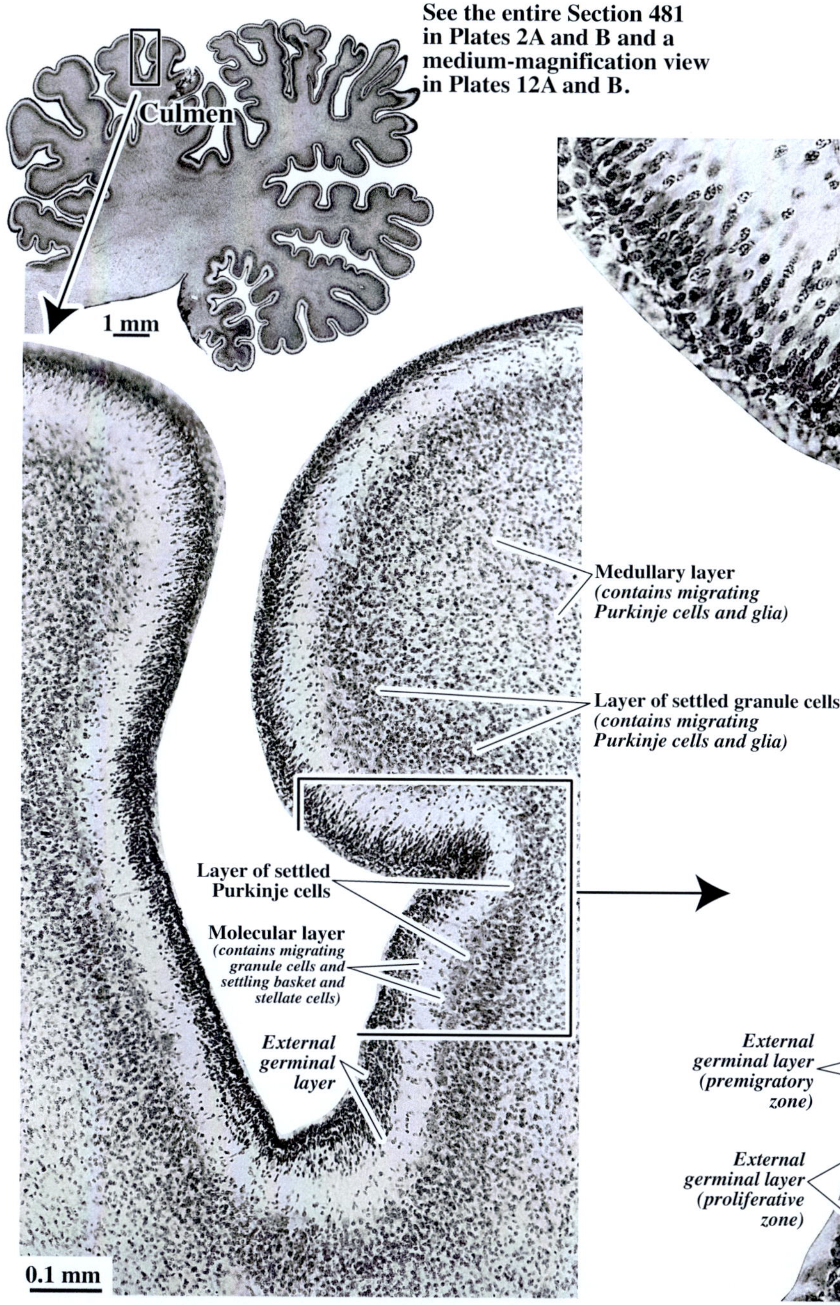

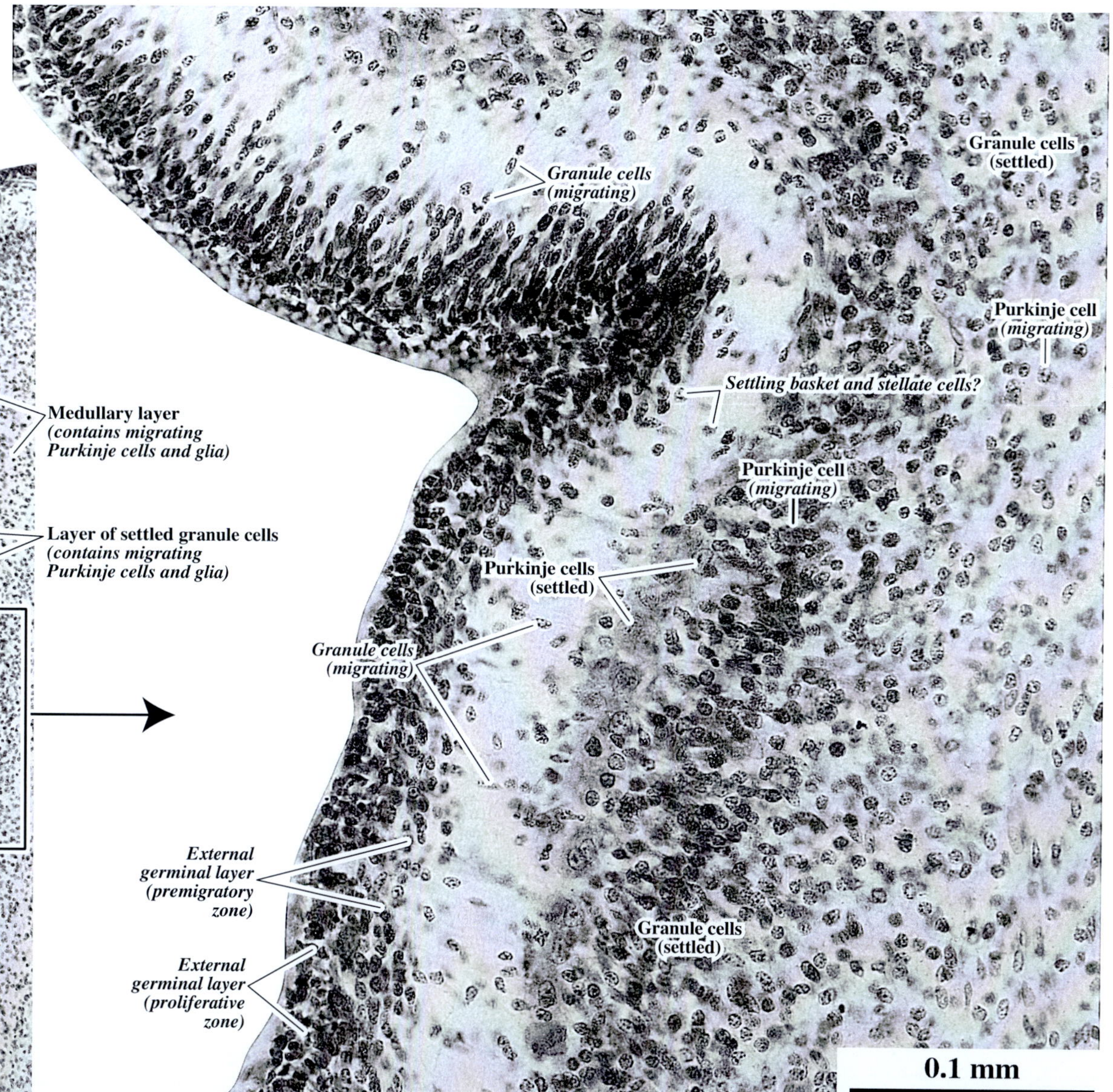

PLATE 23

CR 225 mm, GW 26, Y147-63, Sagittal, Section 481

CEREBELLUM, CENTRAL LOBE, DECLIVE

See the entire Section 481 in Plates 2A and B and a medium-magnification view in Plates 12A and B.

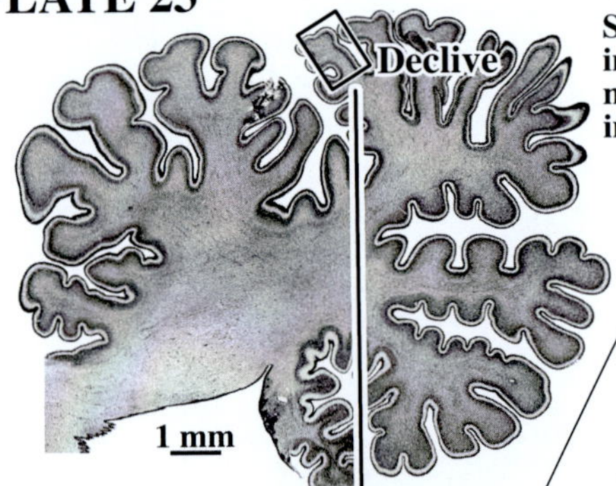

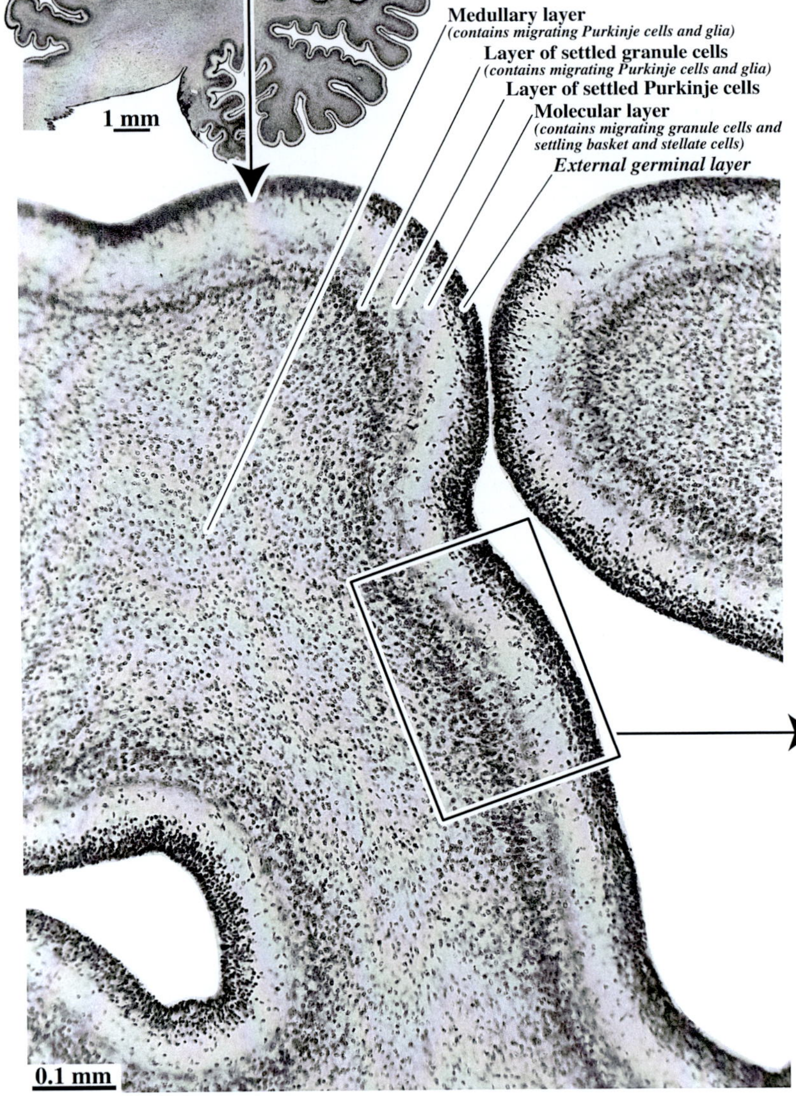

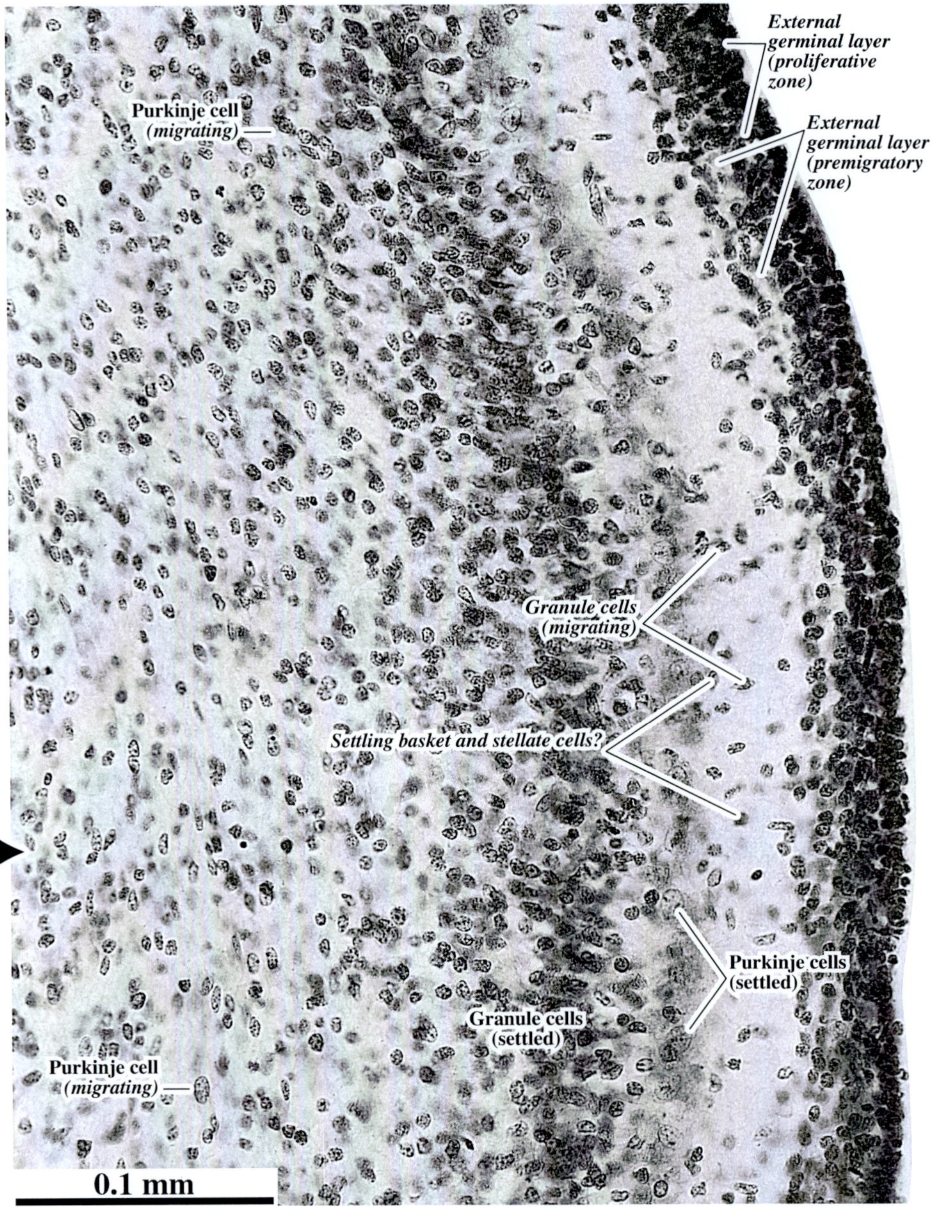

PLATE 24
CR 225 mm, GW 26, Y147-63, Sagittal, Section 481
CEREBELLUM, POSTERIOR LOBE, UVULA

See the entire Section 481 in Plates 2A and B and a medium-magnification view in Plates 12A and B.

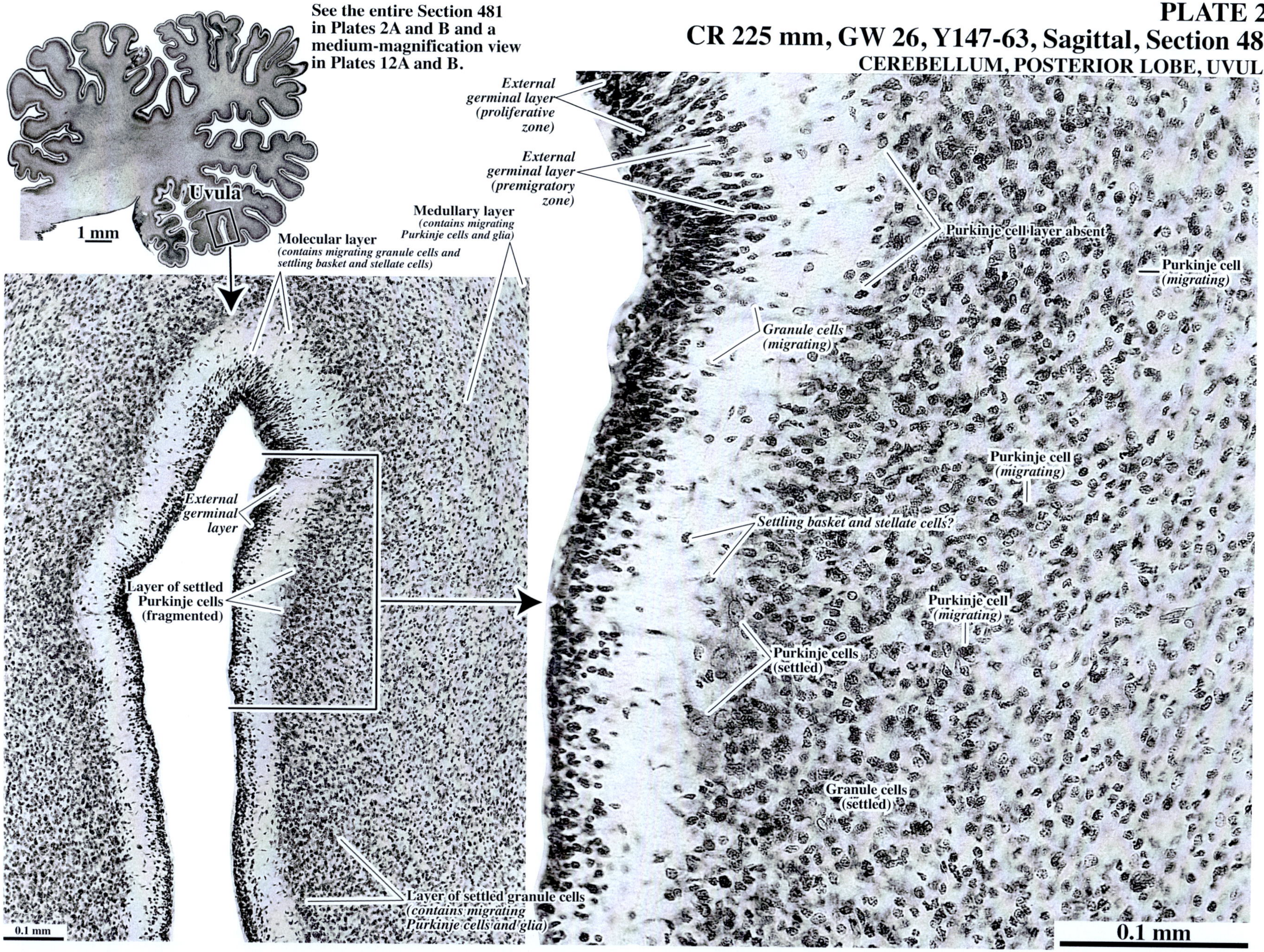

PLATE 25

CR 225 mm, GW 26, Y147-63, Sagittal, Section 481

CEREBELLUM, INFERIOR LOBE, NODULUS

See the entire Section 481 in Plates 2A and B and a medium-magnification view in Plates 12A and B.

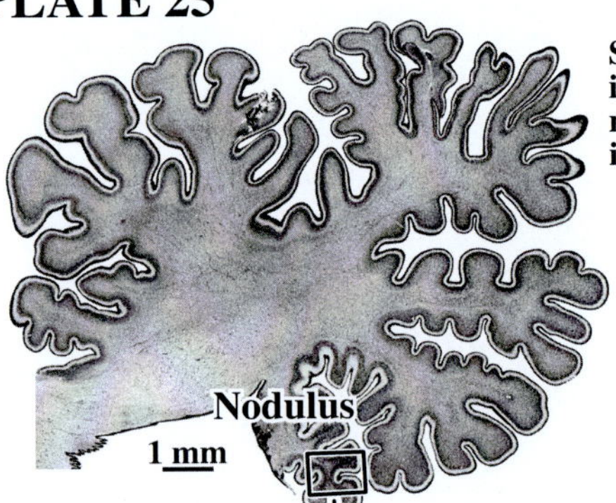

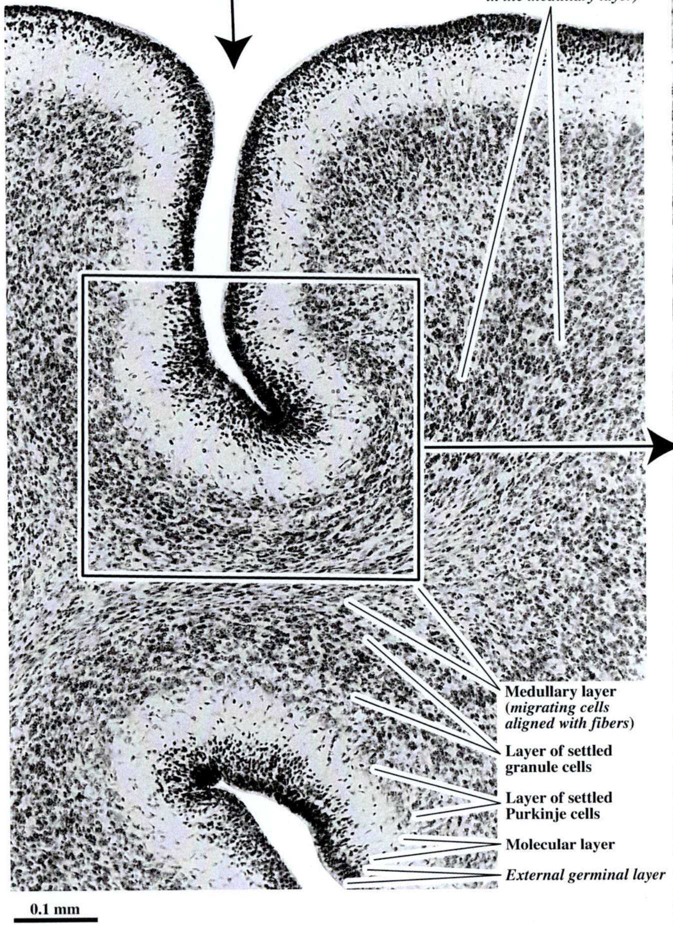

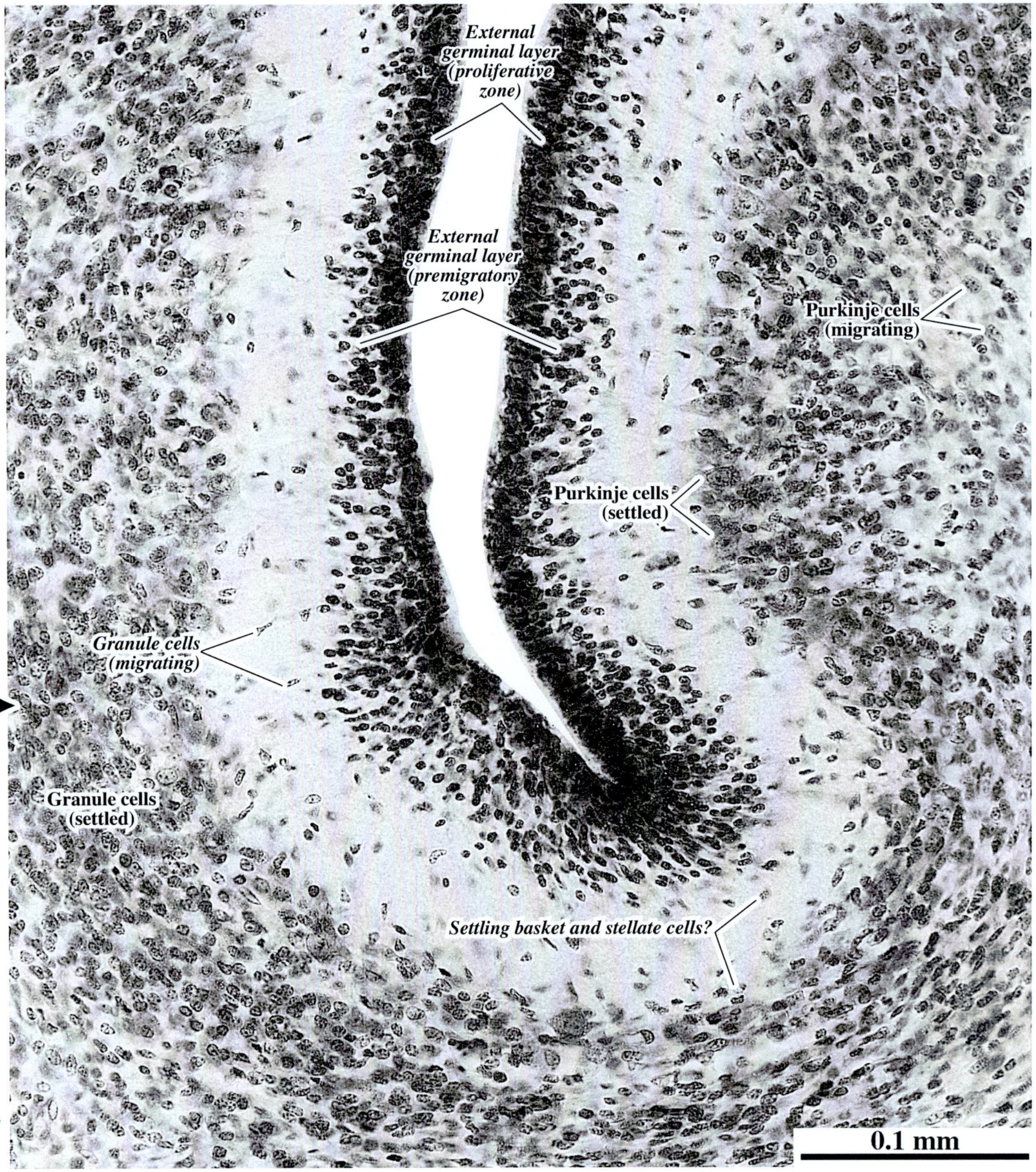

PLATE 26

CR 225 mm, GW 26, Y147-63, Sagittal, Section 481

CEREBELLUM

Germinal Trigone

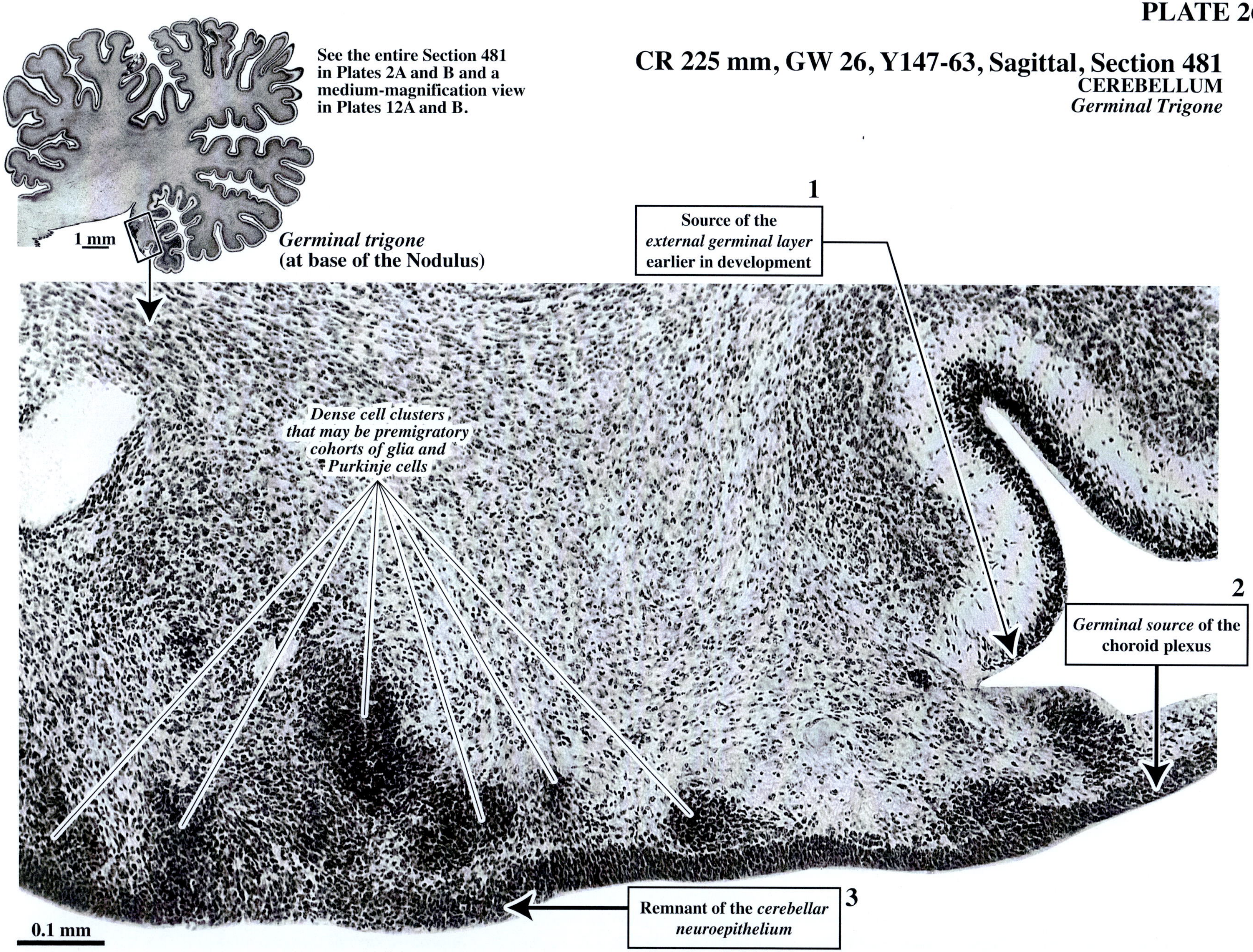

PART III: Y16-59 CR 235 mm (GW 26) Horizontal

This specimen is case number W-16-59 (Perinatal RPSL) in the Yakovlev Collection, a premature stillborn female infant. The brain is classified as a Normative Control in the Yakovlev Collection (Haleem, 1990). It was cut in the horizontal plane in 35-μm and 15-μm thick sections. Since there is no available photograph of this brain before it was embedded and cut, the photograph of the lateral view of another GW 26 brain that Larroche published in 1967 (**Figure 6**) is similar to the features of the brain in Y16-59.

The approximate cutting plane of this brain is indicated in **Figure 7** (facing page) with lines superimposed on the GW26 brain from the Larroche (1967) series. The anterior part of each section (on the left) is dorsal to the posterior part (on the right). As in all other specimens, the sections chosen for illustration are more closely spaced to show small structures in the diencephalon, midbrain, pons, and medulla. Illustrated sections are spaced farther apart when they contain only large brain structures, such as the cerebral cortex, basal ganglia, and cerebellum. Photographs of 15 Nissl-stained low-magnification sections are shown in **Plates 27-41**. **Plates 42-43** show the brainstem at high magnification because these sections are below the cerebral cortex. The core of the brain and the cerebellum are shown at high magnification in **Plates 44-56**.

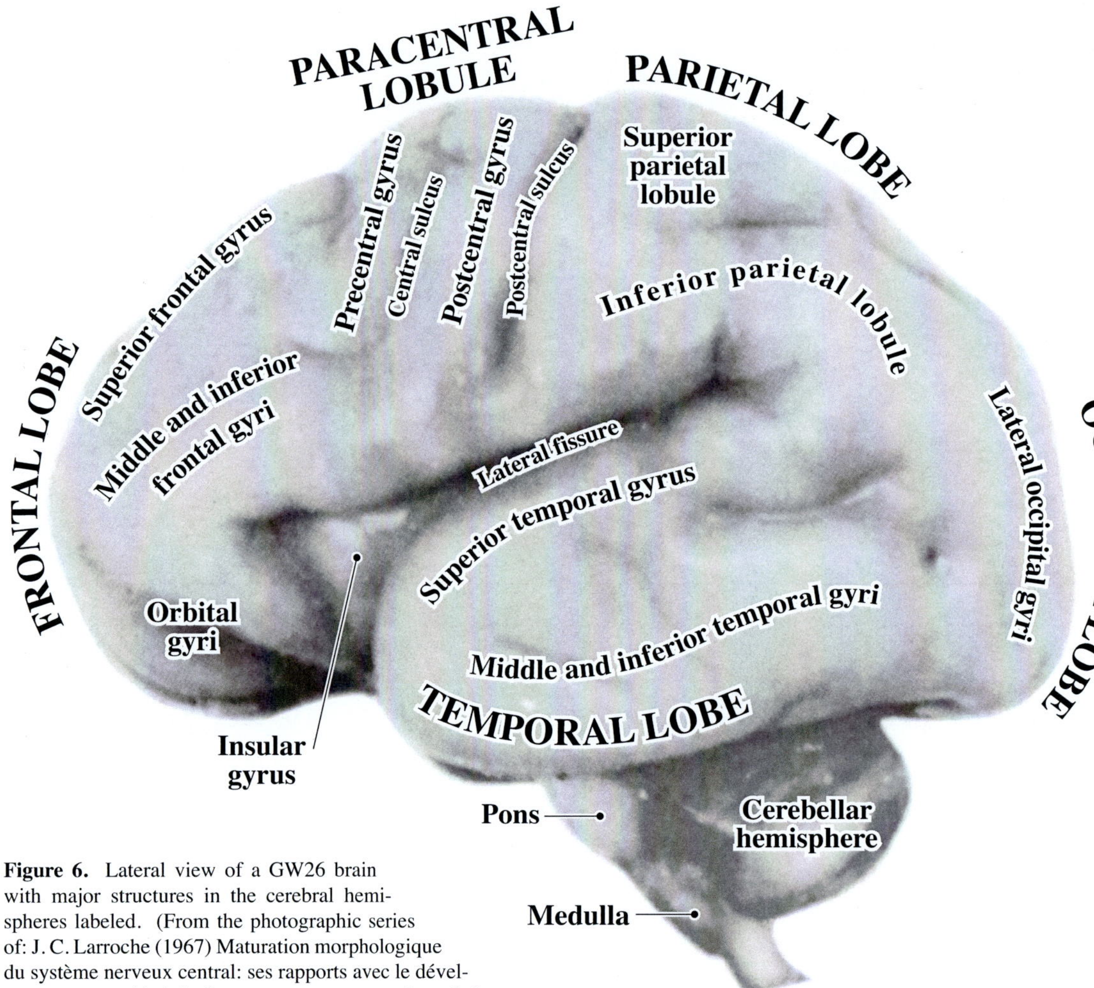

Figure 6. Lateral view of a GW26 brain with major structures in the cerebral hemispheres labeled. (From the photographic series of: J. C. Larroche (1967) Maturation morphologique du système nerveux central: ses rapports avec le développement pondéral du foetus et son age gestationnel. In: *Regional Development of the Brain in Early Life*, A. Minkowski (ed.), London: Blackwell, page 254.)

Y16-59 contains some prominent immature structures. A densely staining ***neuroepithelium/subventricular zone*** is generating neocortical interneurons and glia in all lobes of the cerebral cortex. Remnants of migrating and sojourning neurons and/or glia are also visible in all lobes of the cerebral cortex in ***stratified transitional fields***. Many neurons, glia, and their mitotic precursor cells are still migrating through the olfactory peduncle toward the olfactory bulb (***rostral migratory stream***) from a presumed source area in the germinal matrix at the junction between the cerebral cortex, striatum, and nucleus accumbens. Within the cerebral cortex, definite streams of neurons and glia are in the ***lateral migratory stream*** that percolates through the claustrum, endopiriform nucleus, external capsule, and uncinate fasciculus. These cells appear to be heading toward the insular cortex, primary olfactory cortex, temporal cortex, and basolateral parts of the amygdaloid complex. In the basal ganglia, there is a thick ***neuroepithelium/subventricular zone*** overlying the striatum and nucleus accumbens where neurons are being generated; at least three subdivisions (anteromedial, anterolateral, and posterior) can be distinguished in the striatum. Another region of active neurogenesis in the telencephalon is the ***subgranular zone*** in the hilus of the dentate gyrus that is the source of granule cells. Other structures in the telencephalon, such as the septum, fornix, and Ammon's horn part of the hippocampus, have only a thin, darkly staining layer at the ventricle, and these are presumed to be generating glia, cells of the choroid plexus, and the ependymal lining of the ventricle.

Most of the structures in the diencephalon appear to be settled and are maturing. The third ventricle is lined by a ***glioepithelium/ependyma***. A convoluted ***glioepithelium/ependyma*** lines the cerebral aqueduct in the midbrain that continues into the anterior fourth ventricle. A smooth ***glioepithelium/ependyma*** lines the posterior fourth ventricle through the remainder of the pons and medulla. A convoluted ***glioepithelium/ependyma*** lines the floor of the fourth ventricle throughout the medulla.In the lower medulla and spinal cord, this specimen is remarkable for showing dense myelination gliosis in the cuneate fasciculus.

The ***external germinal layer*** covering the cerebellar cortex is actively producing basket, stellate, and granule cells. It is one prong of the ***germinal trigone*** at the base of the nodulus and along the floccular peduncle. Another prong is choroid plexus stem cells. The third prong is the ***glioepithelium/ependym*** beneath the cerebellum.

GW 26 HORIZONTAL SECTION PLANES

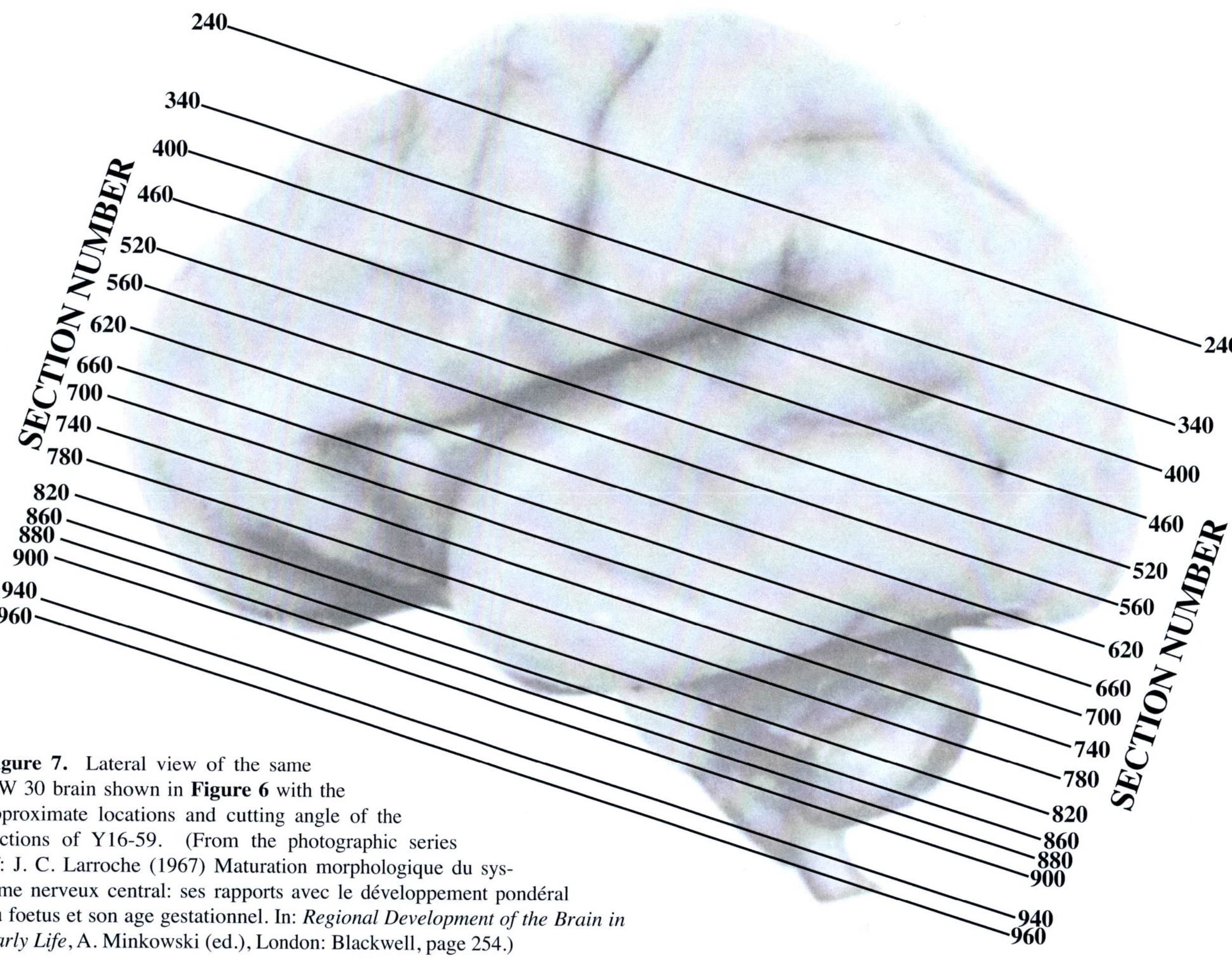

Figure 7. Lateral view of the same GW 30 brain shown in **Figure 6** with the approximate locations and cutting angle of the sections of Y16-59. (From the photographic series of: J. C. Larroche (1967) Maturation morphologique du système nerveux central: ses rapports avec le développement pondéral du foetus et son age gestationnel. In: *Regional Development of the Brain in Early Life*, A. Minkowski (ed.), London: Blackwell, page 254.)

PLATE 27A
CR 235 mm
GW 26, Y16-59
Horizontal
Section 240

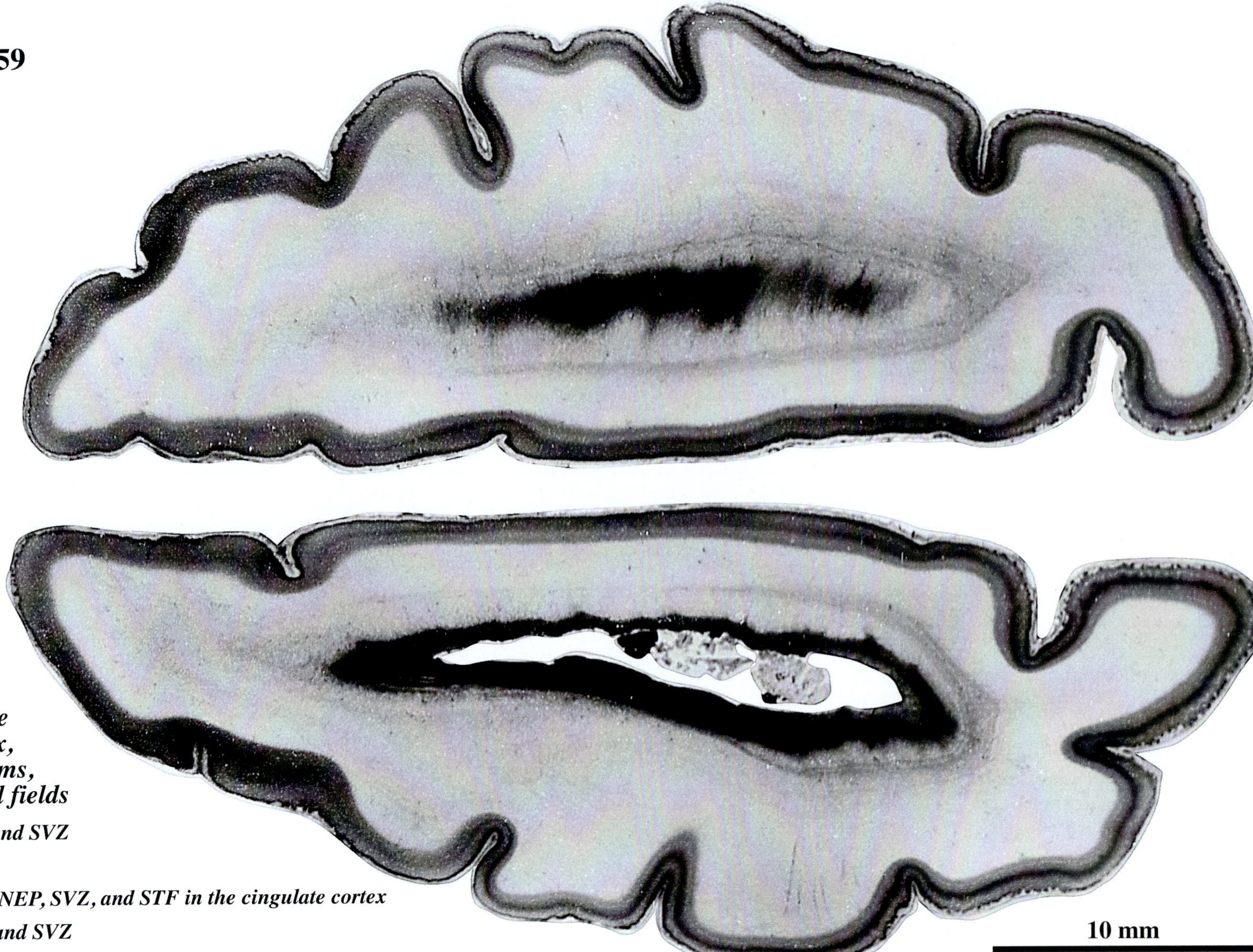

Remnants of the germinal matrix, migratory streams, and transitional fields

1 *Frontal NEP and SVZ*
2 *Frontal STF*
3 *Callosal GEP, NEP, SVZ, and STF in the cingulate cortex*
4 *Parietal NEP and SVZ*
5 *Parietal STF*
6 *Paracentral NEP and SVZ*
7 *Paracentral STF*
8 *Subpial granular layer (cortical)*

GEP - Glioepithelium
NEP - Neuroepithelium
STF - Stratified transitional field
SVZ - Subventricular zone

PLATE 27B

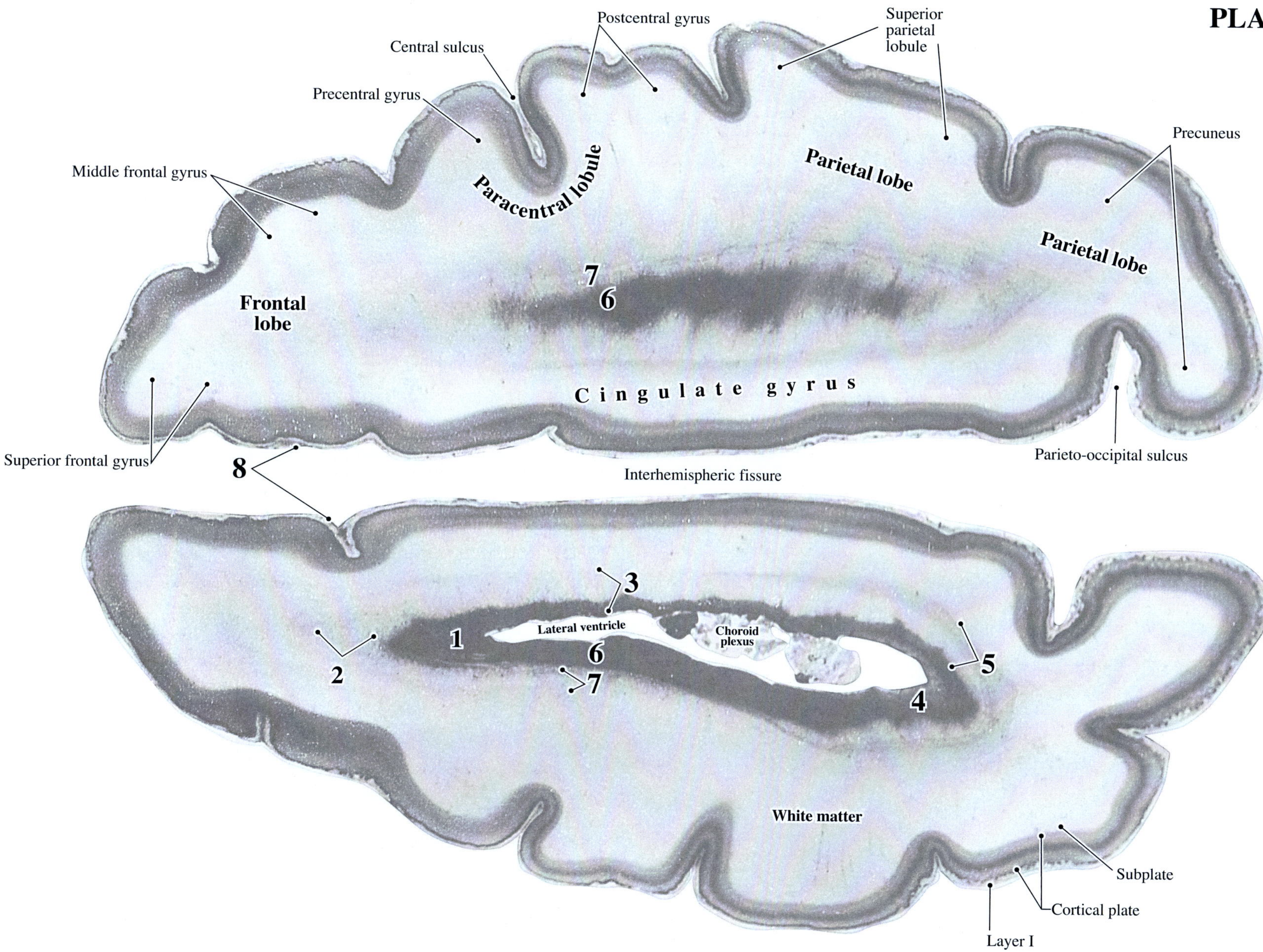
Postcentral gyrus
Superior parietal lobule
Central sulcus
Precentral gyrus
Middle frontal gyrus
Paracentral lobule
Parietal lobe
Precuneus
Parietal lobe
7
6
Frontal lobe
Cingulate gyrus
Superior frontal gyrus
8
Parieto-occipital sulcus
Interhemispheric fissure
3
Lateral ventricle
Choroid plexus
1
6
2
7
5
4
White matter
Subplate
Cortical plate
Layer I

PLATE 28A
CR 235 mm
GW 26, Y16-59
Horizontal
Section 340

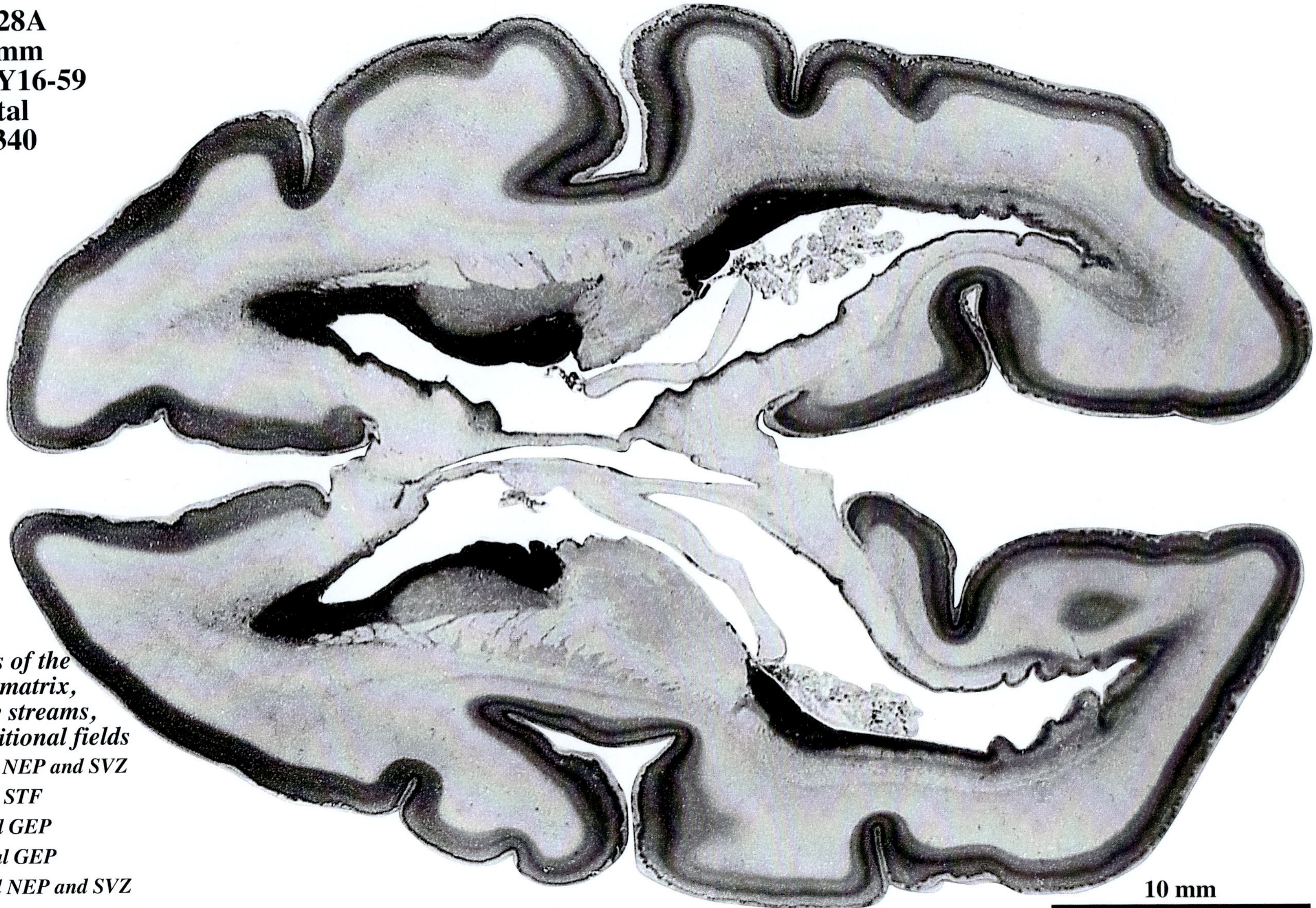

Remnants of the germinal matrix, migratory streams, and transitional fields

1 *Frontal NEP and SVZ*
2 *Frontal STF*
3 *Callosal GEP*
4 *Fornical GEP*
5 *Parietal NEP and SVZ*
6 *Parietal STF*
7 *Posterior striatalNEP and SVZ*
8 *Anterolateral striatal NEP and SVZ*
9 *Anteromedial striatal NEP and SVZ*
10 *Strionuclear GEP*
11 *Subpial granular layer (cortical)*

GEP - Glioepithelium
NEP - Neuroepithelium
STF - Stratified transitional field
SVZ - Subventricular zone

PLATE 28B

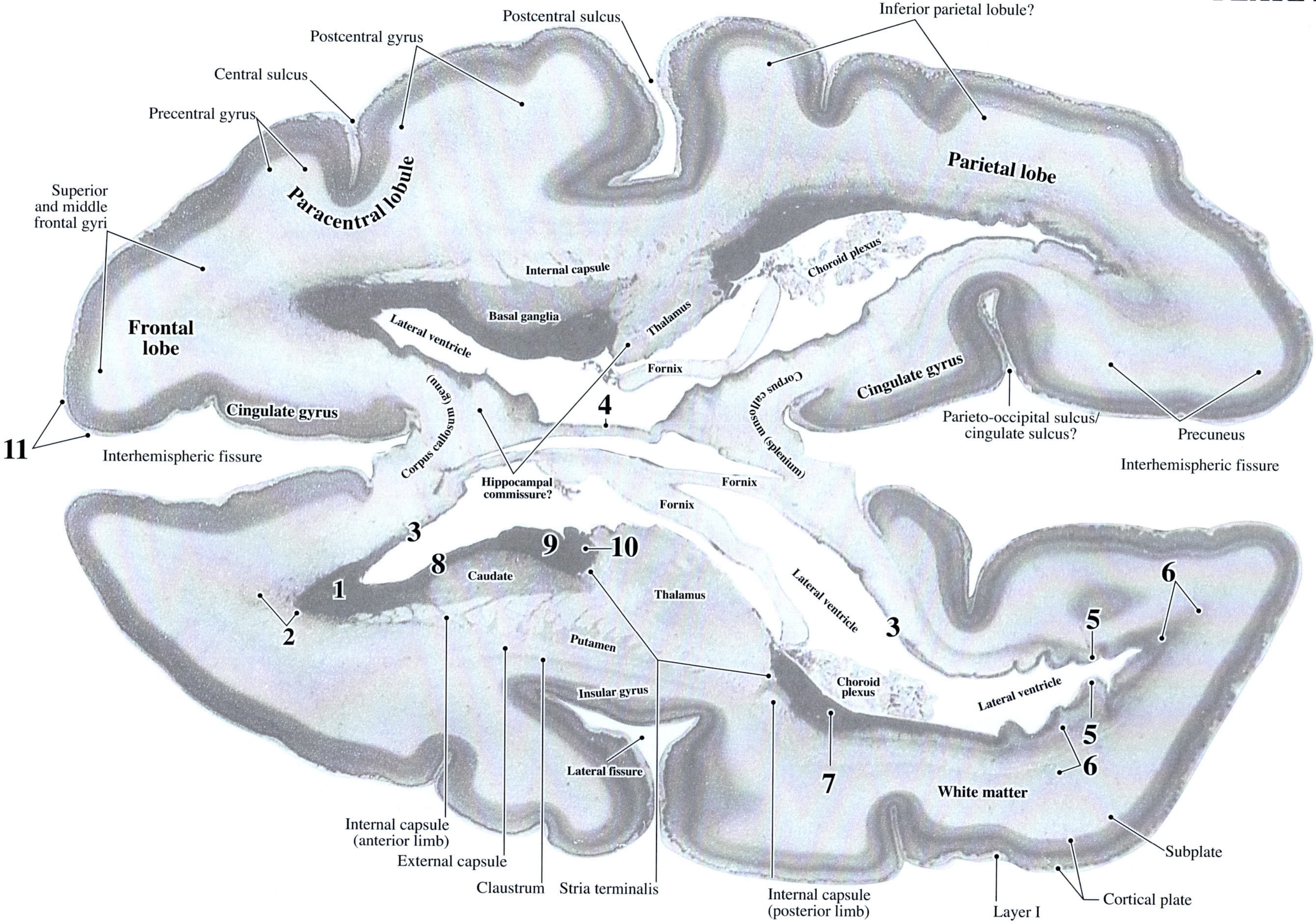
Postcentral sulcus
Inferior parietal lobule?
Postcentral gyrus
Central sulcus
Precentral gyrus
Superior and middle frontal gyri
Paracentral lobule
Parietal lobe
Internal capsule
Choroid plexus
Basal ganglia
Thalamus
Frontal lobe
Lateral ventricle
Fornix
Corpus callosum (genu)
Corpus callosum (splenium)
Cingulate gyrus
Cingulate gyrus
4
Parieto-occipital sulcus/ cingulate sulcus?
Precuneus
11
Interhemispheric fissure
Interhemispheric fissure
Hippocampal commissure?
Fornix
Fornix
3
9
10
8
Caudate
1
Thalamus
Lateral ventricle
6
2
5
3
Putamen
Choroid plexus
Lateral ventricle
Insular gyrus
5
6
Lateral fissure
7
White matter
Subplate
Internal capsule (anterior limb)
External capsule
Claustrum
Stria terminalis
Internal capsule (posterior limb)
Layer I
Cortical plate

PLATE 29A
CR 235 mm
GW 26, Y16-59
Horizontal
Section 400

See detail of the brain core in Plates 44A and B.

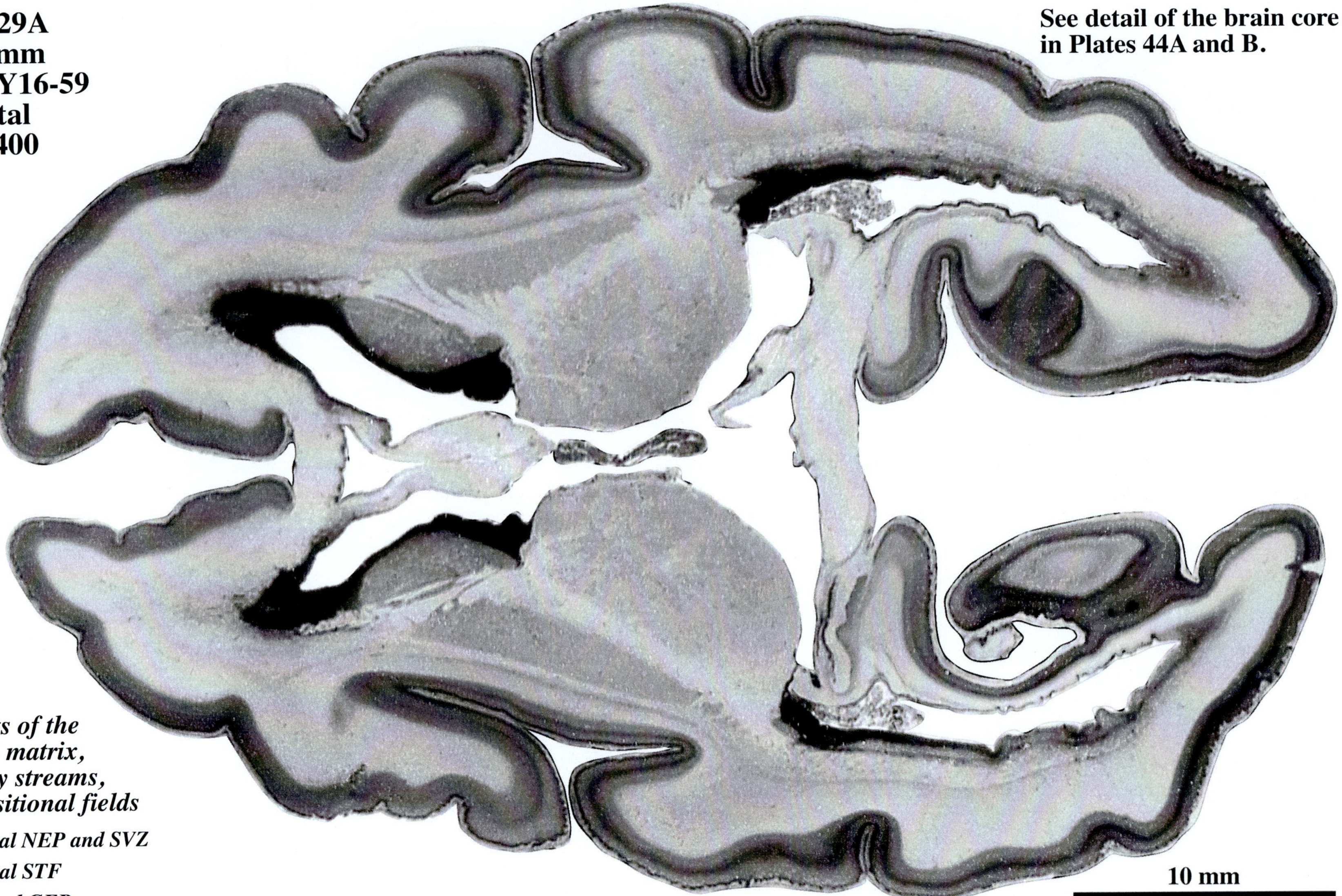

Remnants of the germinal matrix, migratory streams, and transitional fields

1 *Frontal NEP and SVZ*
2 *Frontal STF*
3 *Callosal GEP*
4 *Callosal sling*
5 *Fornical GEP*
6 *Parahippocampal NEP, SVZ, and STF*
7 *Occipital NEP and SVZ*
8 *Occipital STF*
9 *Parietal NEP and SVZ*
10 *Parietal STF*
11 *Posterior striatal NEP and SVZ*
12 *Anterolateral striatal NEP and SVZ*
13 *Anteromedial striatal NEP and SVZ*
14 *Strionuclear GE*
15 *Subpial granular layer (cortical)*

GEP - Glioepithelium
NEP - Neuroepithelium
STF - Stratified transitional field
SVZ - Subventricular zone

PLATE 29B

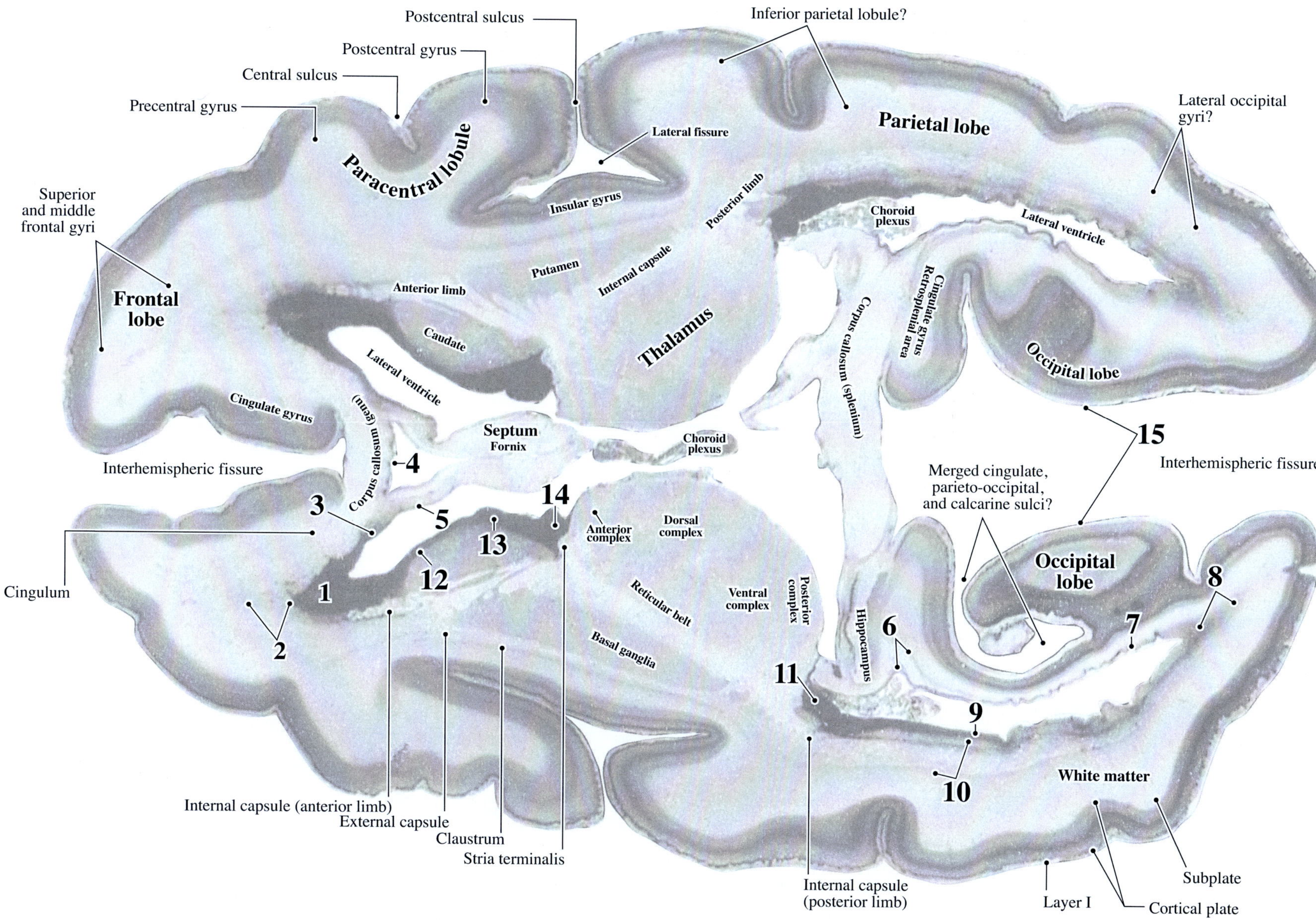

Postcentral sulcus
Inferior parietal lobule?
Postcentral gyrus
Central sulcus
Precentral gyrus
Lateral fissure
Parietal lobe
Lateral occipital gyri?
Paracentral lobule
Insular gyrus
Posterior limb
Choroid plexus
Lateral ventricle
Superior and middle frontal gyri
Putamen
Internal capsule
Frontal lobe
Anterior limb
Cingulate gyrus Retrosplenial area
Corpus callosum (splenium)
Caudate
Thalamus
Lateral ventricle
Occipital lobe
Cingulate gyrus
Corpus callosum (genu)
Septum
Fornix
Choroid plexus
15
4
Interhemispheric fissure
Interhemispheric fissure
Merged cingulate, parieto-occipital, and calcarine sulci?
3
14
5
Anterior complex
Dorsal complex
13
Cingulum
12
Occipital lobe
8
1
Reticular belt
Ventral complex
Posterior complex
7
Hippocampus
6
2
Basal ganglia
11
9
10
White matter
Internal capsule (anterior limb)
External capsule
Claustrum
Stria terminalis
Internal capsule (posterior limb)
Subplate
Layer I
Cortical plate

PLATE 30A
CR 235 mm
GW 26, Y16-59
Horizontal
Section 460

See detail of the brain core in Plates 45A and B.

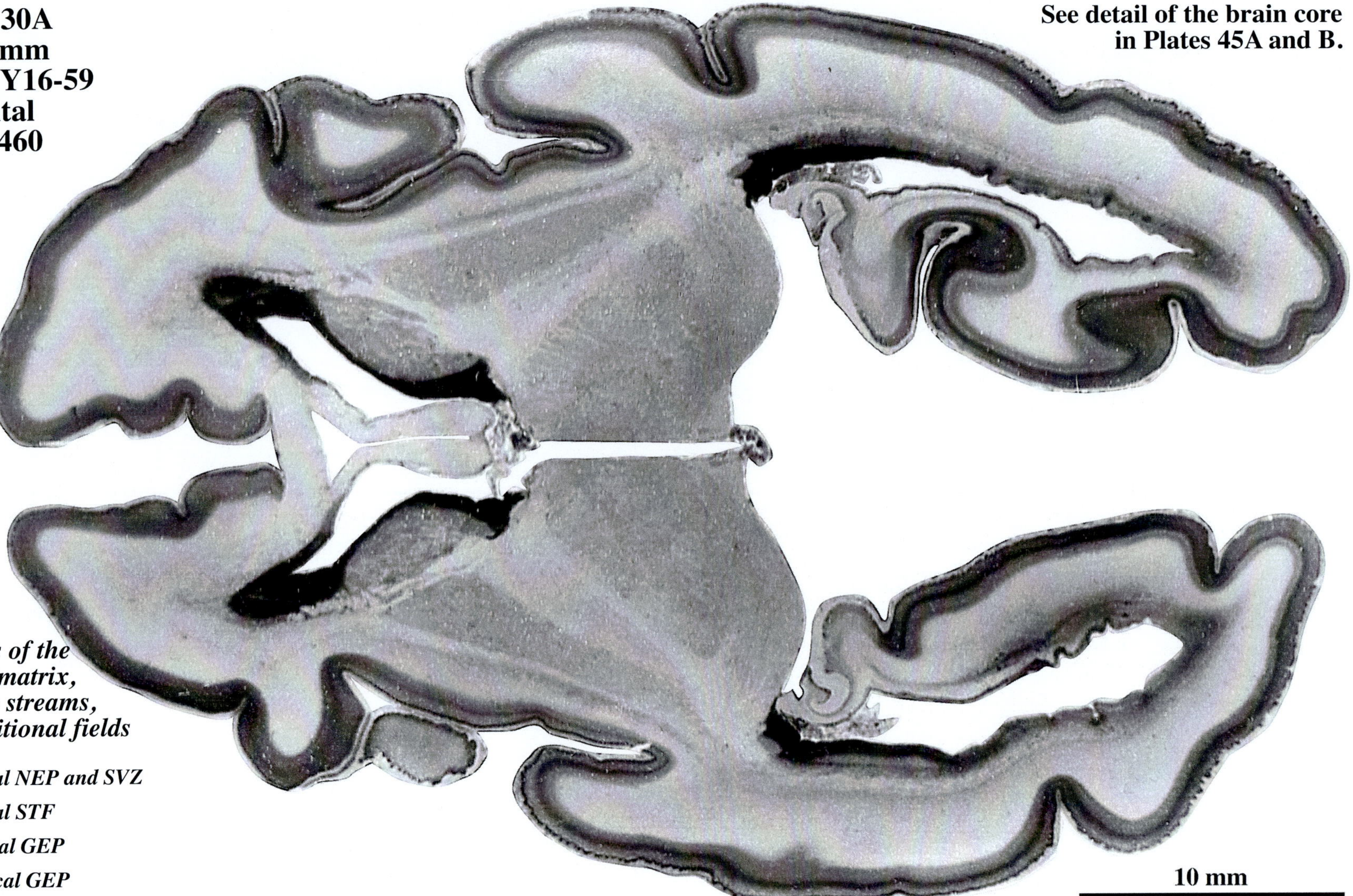

Remnants of the germinal matrix, migratory streams, and transitional fields

1 *Frontal NEP and SVZ*
2 *Frontal STF*
3 *Callosal GEP*
4 *Fornical GEP*
5 *Parahippocampal NEP, SVZ, and STF*
6 *Occipital NEP and SVZ*
7 *Occipital STF*
8 *Temporal NEP and SVZ*
9 *Temporal STF*
10 *Alvear GEP*
11 *Subgranular zone (dentate)*
12 *Posterior striatal NEP and SVZ*
13 *Anterolateral striatal NEP and SVZ*
14 *Anteromedial striatal NEP and SVZ*
15 *StrionuclearGEP*
16 *Subpial granular layer (cortical)*

10 mm

GEP - Glioepithelium
NEP - Neuroepithelium
STF - Stratified transitional field
SVZ - Subventricular zone

PLATE 30B

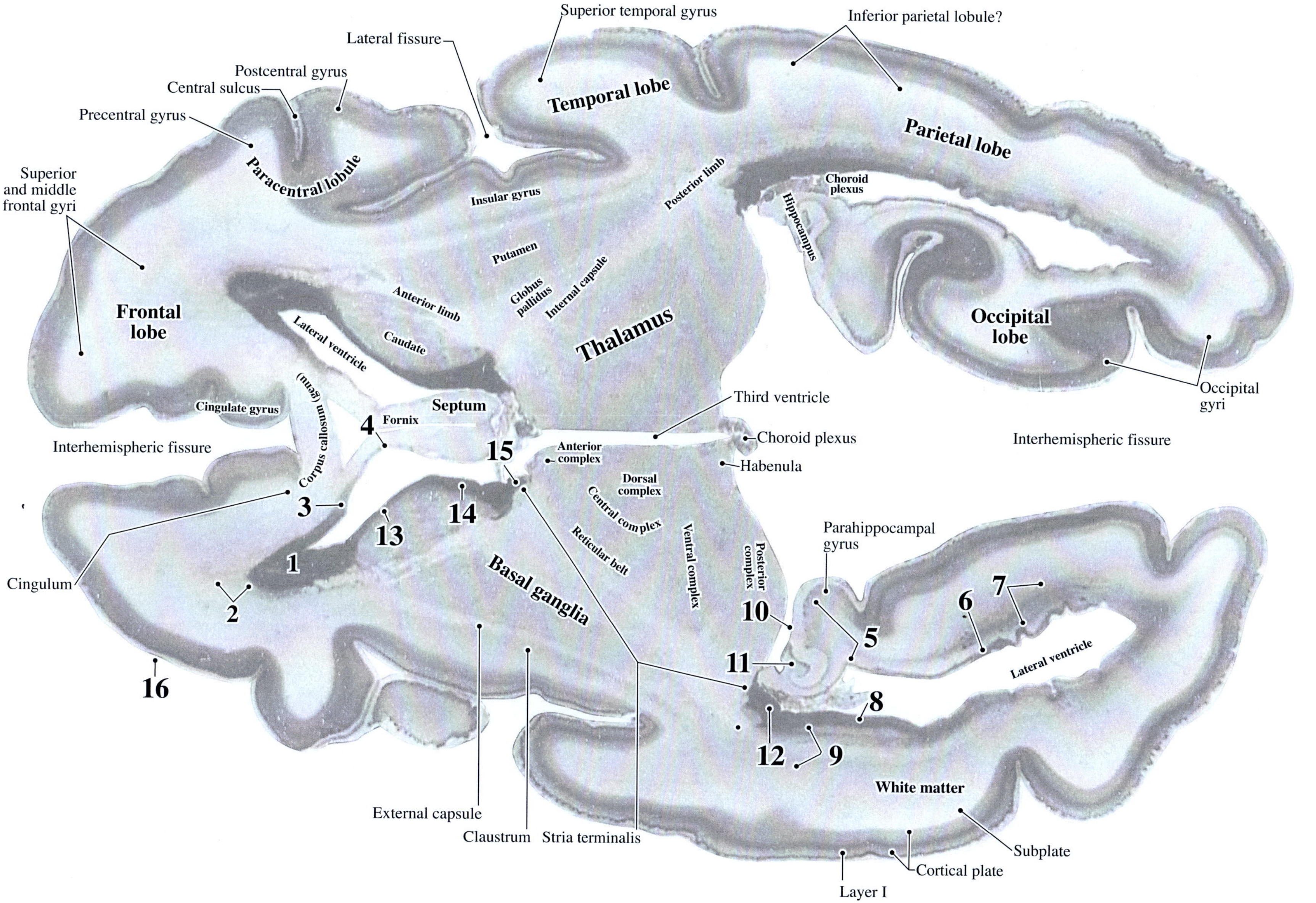
Superior temporal gyrus
Inferior parietal lobule?
Lateral fissure
Postcentral gyrus
Central sulcus
Precentral gyrus
Temporal lobe
Parietal lobe
Superior and middle frontal gyri
Paracentral lobule
Insular gyrus
Posterior limb
Choroid plexus
Hippocampus
Putamen
Globus pallidus
Internal capsule
Anterior limb
Frontal lobe
Lateral ventricle
Caudate
Thalamus
Occipital lobe
Corpus callosum (genu)
Cingulate gyrus
Septum
Fornix
Third ventricle
Occipital gyri
Interhemispheric fissure
Choroid plexus
Interhemispheric fissure
Anterior complex
Habenula
Dorsal complex
Central complex
Reticular belt
Ventral complex
Posterior complex
Parahippocampal gyrus
Basal ganglia
Cingulum
Lateral ventricle
White matter
External capsule
Claustrum
Stria terminalis
Subplate
Cortical plate
Layer I
1
2
3
4
5
6
7
8
9
10
11
12
13
14
15
16

PLATE 31A
CR 235 mm
GW 26, Y16-59
Horizontal
Section 520

See detail of the brain core and cerebellum in Plates 46A and B.

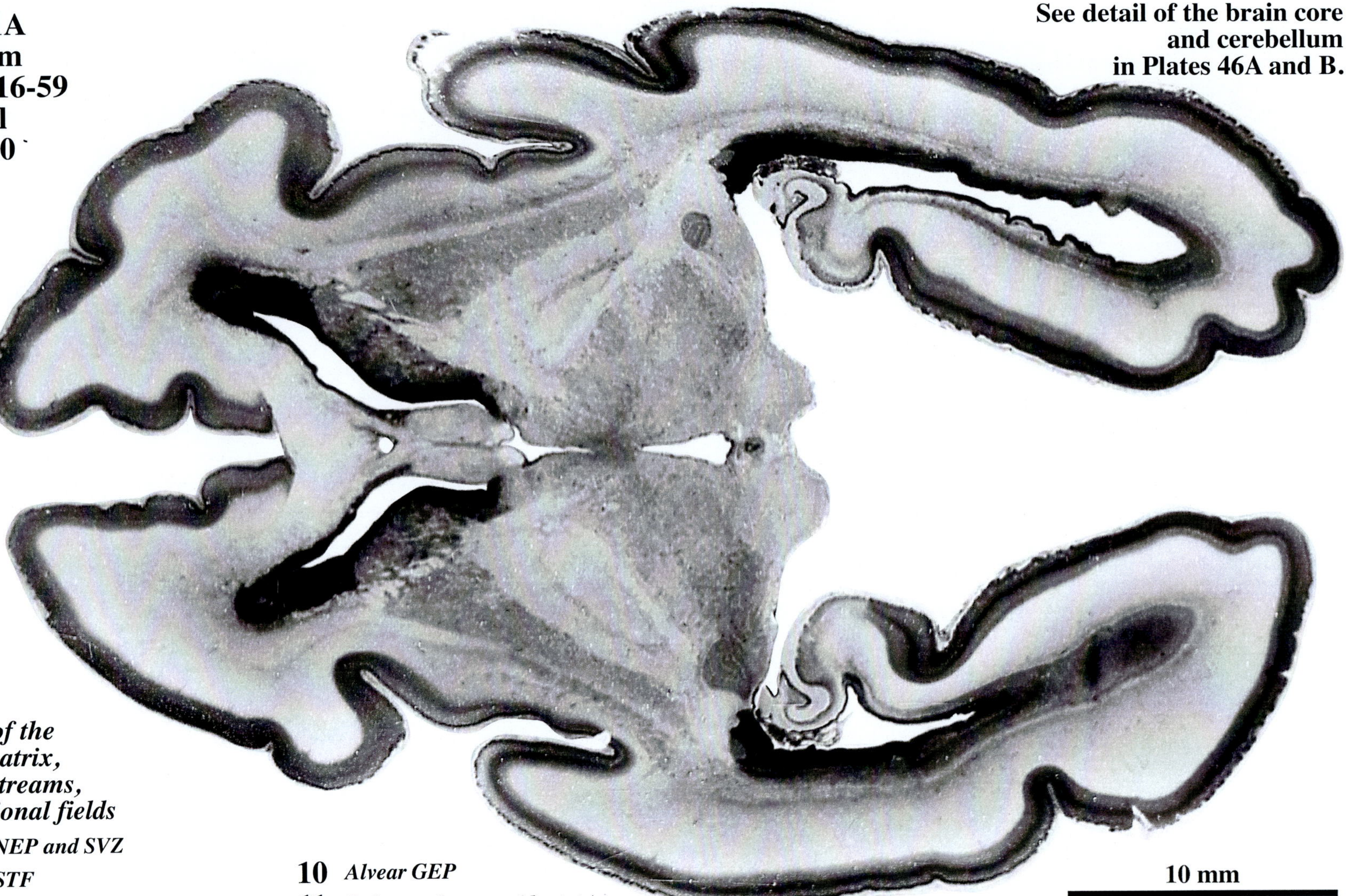

Remnants of the germinal matrix, migratory streams, and transitional fields

1 *Frontal NEP and SVZ*
2 *Frontal STF*
3 *Callosal GEP*
4 *Fornical GEP*
5 *Parahippocampal NEP, SVZ, and STF*
6 *Occipital NEP and SVZ*
7 *Occipital STF*
8 *Temporal NEP and SVZ*
9 *Temporal STF*
10 *Alvear GEP*
11 *Subgranular zone (dentate)*
12 *Lateral migratory stream (cortical)*
13 *Posterior striatal NEP and SVZ*
14 *Anterolateral striatal NEP and SVZ*
15 *Anteromedial striatal NEP and SVZ*
16 *Strionuclear GEP*
17 *Subpial granular layer (cortical)*

GEP - Glioepithelium
NEP - Neuroepithelium
STF - Stratified transitional field
SVZ - Subventricular zone

PLATE 31B

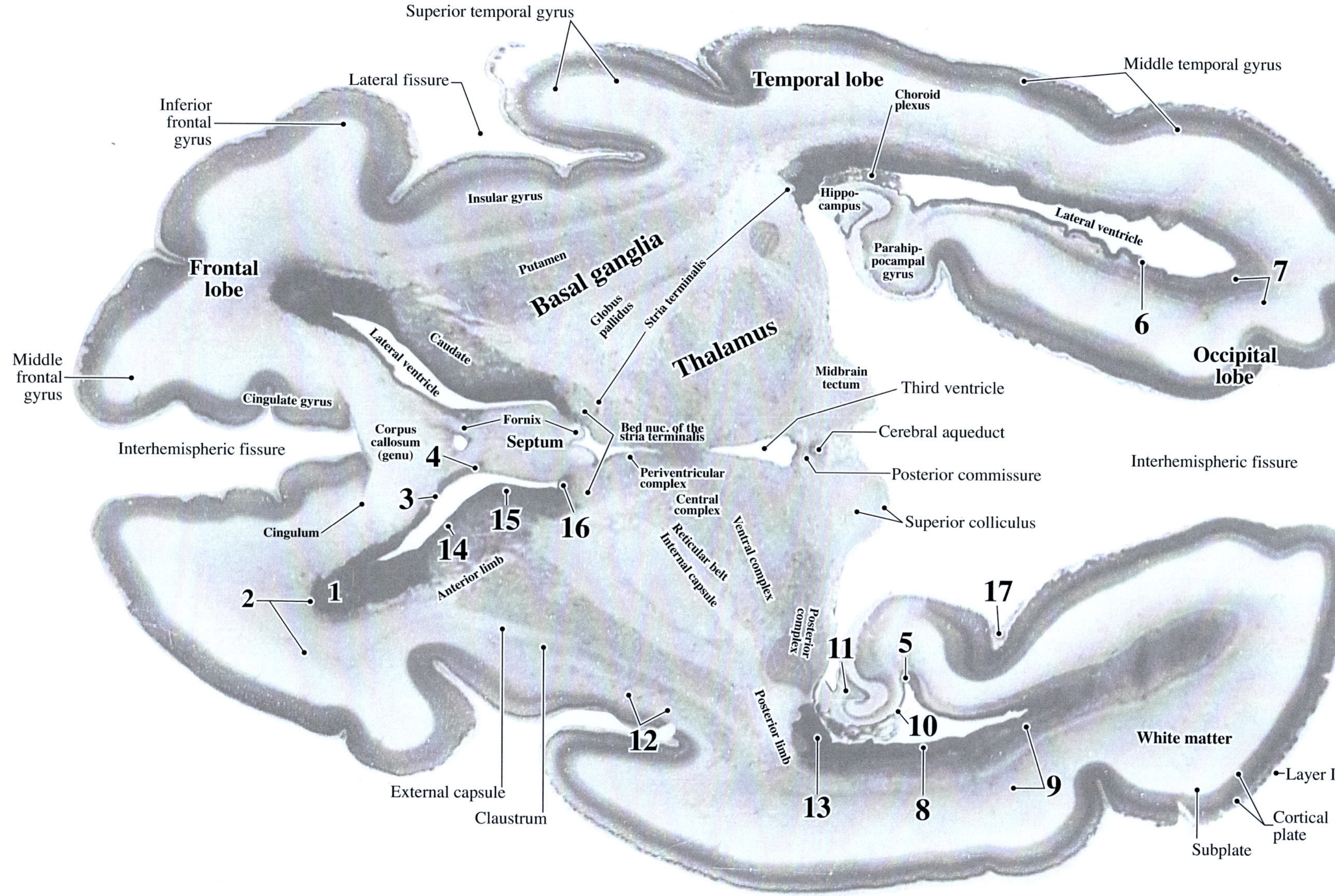
Superior temporal gyrus
Lateral fissure
Inferior frontal gyrus
Temporal lobe
Choroid plexus
Middle temporal gyrus
Insular gyrus
Hippo-campus
Lateral ventricle
Parahip-pocampal gyrus
Putamen
Basal ganglia
Frontal lobe
Globus pallidus
Stria terminalis
7
6
Caudate
Lateral ventricle
Thalamus
Occipital lobe
Middle frontal gyrus
Midbrain tectum
Third ventricle
Cingulate gyrus
Fornix
Bed nuc. of the stria terminalis
Cerebral aqueduct
Corpus callosum (genu)
Septum
Interhemispheric fissure
4
Posterior commissure
Interhemispheric fissure
Periventricular complex
3
Central complex
15
16
Superior colliculus
Cingulum
Reticular belt
Ventral complex
14
Internal capsule
Anterior limb
17
2
1
Posterior complex
11
5
10
12
Posterior limb
White matter
Layer I
9
External capsule
13
8
Cortical plate
Claustrum
Subplate

PLATE 32A
CR 235 mm
GW 26, Y16-59
Horizontal
Section 560

See detail of the brain core and cerebellum in Plates 47A and B.

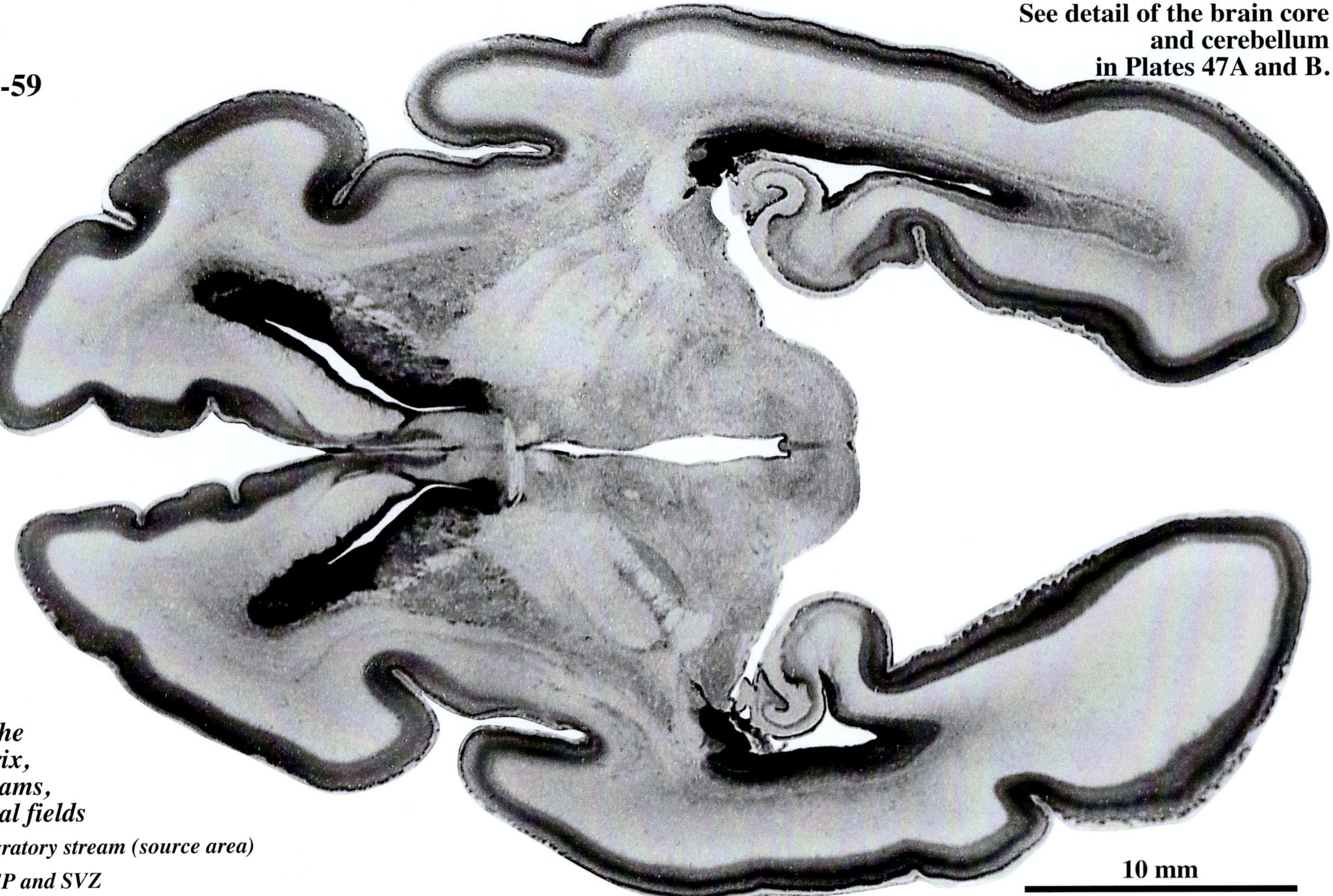

Remnants of the germinal matrix, migratory streams, and transitional fields

1 *Rostral migratory stream (source area)*
2 *Frontal NEP and SVZ*
3 *Frontal STF*
4 *Callosal GEP*
5 *Parahippocampal NEP, SVZ, and STF*
6 *Occipital STF*
7 *Temporal NEP and SVZ*
8 *Temporal STF*
9 *Alvear GEP*
10 *Subgranular zone (dentate)*
11 *Lateral migratory stream (cortical)*
12 *Posterior striatal NEP and SVZ*
13 *Anterolateral striatal NEP and SVZ*
14 *Anteromedial striatal NEP and SVZ*
15 *Strionuclear GEP*
16 *Subpial granular layer (cortical)*

GEP - Glioepithelium
NEP - Neuroepithelium
STF - Stratified transitional field
SVZ - Subventricular zone

PLATE 32B

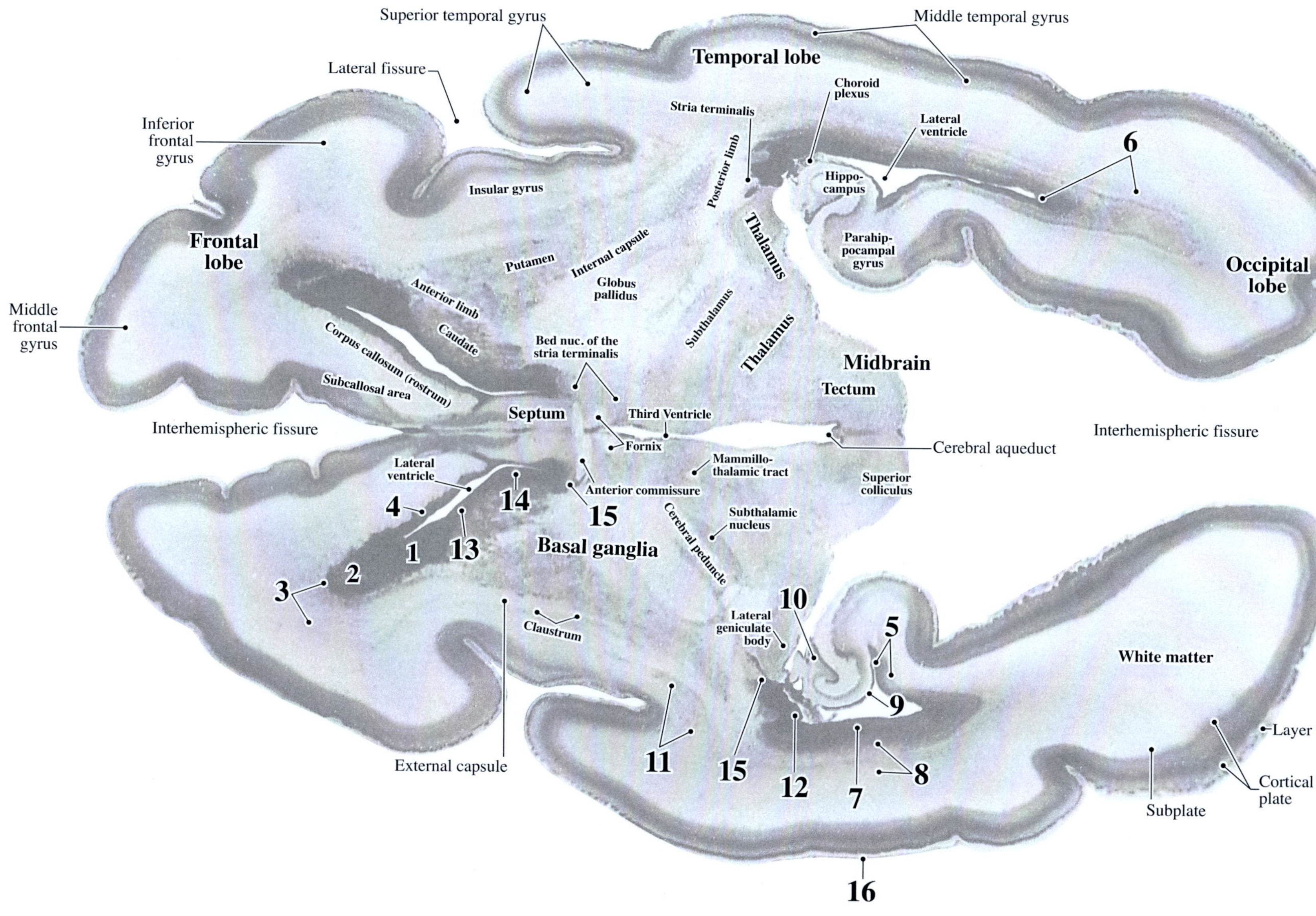
Superior temporal gyrus
Middle temporal gyrus
Temporal lobe
Lateral fissure
Choroid plexus
Stria terminalis
Inferior frontal gyrus
Lateral ventricle
6
Posterior limb
Hippocampus
Insular gyrus
Thalamus
Parahippocampal gyrus
Frontal lobe
Internal capsule
Putamen
Globus pallidus
Occipital lobe
Anterior limb
Middle frontal gyrus
Caudate
Subthalamus
Corpus callosum (rostrum)
Bed nuc. of the stria terminalis
Thalamus
Midbrain
Tectum
Subcallosal area
Septum
Third Ventricle
Interhemispheric fissure
Fornix
Cerebral aqueduct
Interhemispheric fissure
Lateral ventricle
Mammillo-thalamic tract
Anterior commissure
Superior colliculus
14
4
15
Subthalamic nucleus
Cerebral peduncle
1
13
Basal ganglia
2
3
10
Lateral geniculate body
5
Claustrum
White matter
9
Layer I
11
15
12
7
8
Cortical plate
External capsule
Subplate
16

PLATE 33A
CR 235 mm
GW 26, Y16-59
Horizontal
Section 620

See detail of the brain core and cerebellum in Plates 48A and B.

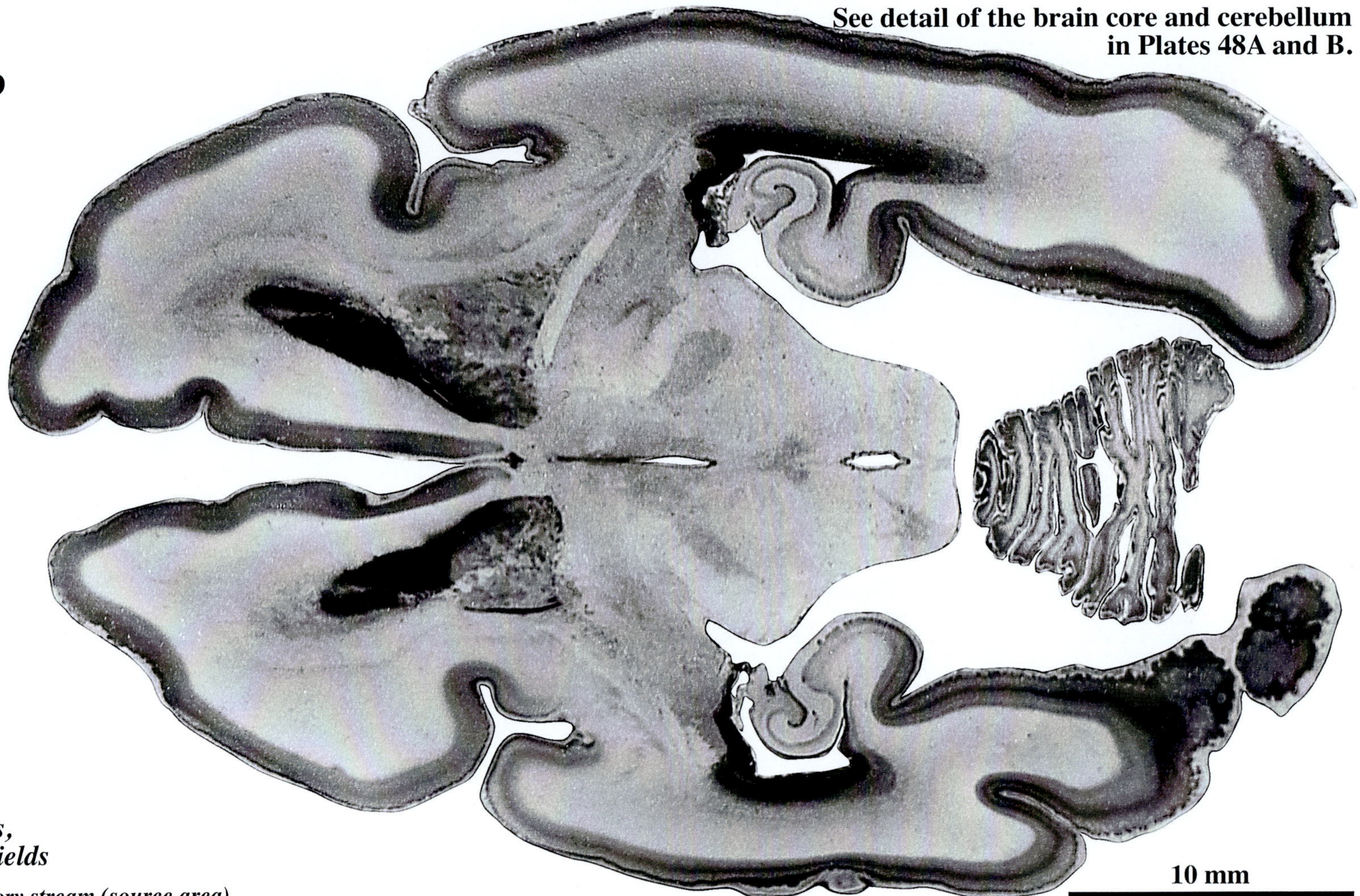

Remnants of the germinal matrix, migratory streams, and transitional fields

1 *Rostral migratory stream (source area)*
2 *Frontal NEP and SVZ*
3 *Frontal STF*
4 *Parahippocampal NEP, SVZ, and STF*
5 *Temporal NEP and SVZ*
6 *Temporal STF*
7 *Alvear GEP*
8 *Subgranular zone (dentate)*
9 *Lateral migratory stream (cortical)*
10 *Posterior striatal NEP and SVZ*
11 *Accumbent NEP (intermingled with the rostral migratory stream)*
12 *Diencephalic (hypothalamic) G/EP*
13 *Mesencephalic G/EP*
14 *External germinal layer (cerebellar)*
15 *Subpial granular layer (cortical)*

GEP - Glioepithelium
G/EP - Glioepithelium/ependyma
NEP - Neuroepithelium
STF - Stratified transitional field
SVZ - Subventricular zone

PLATE 33B

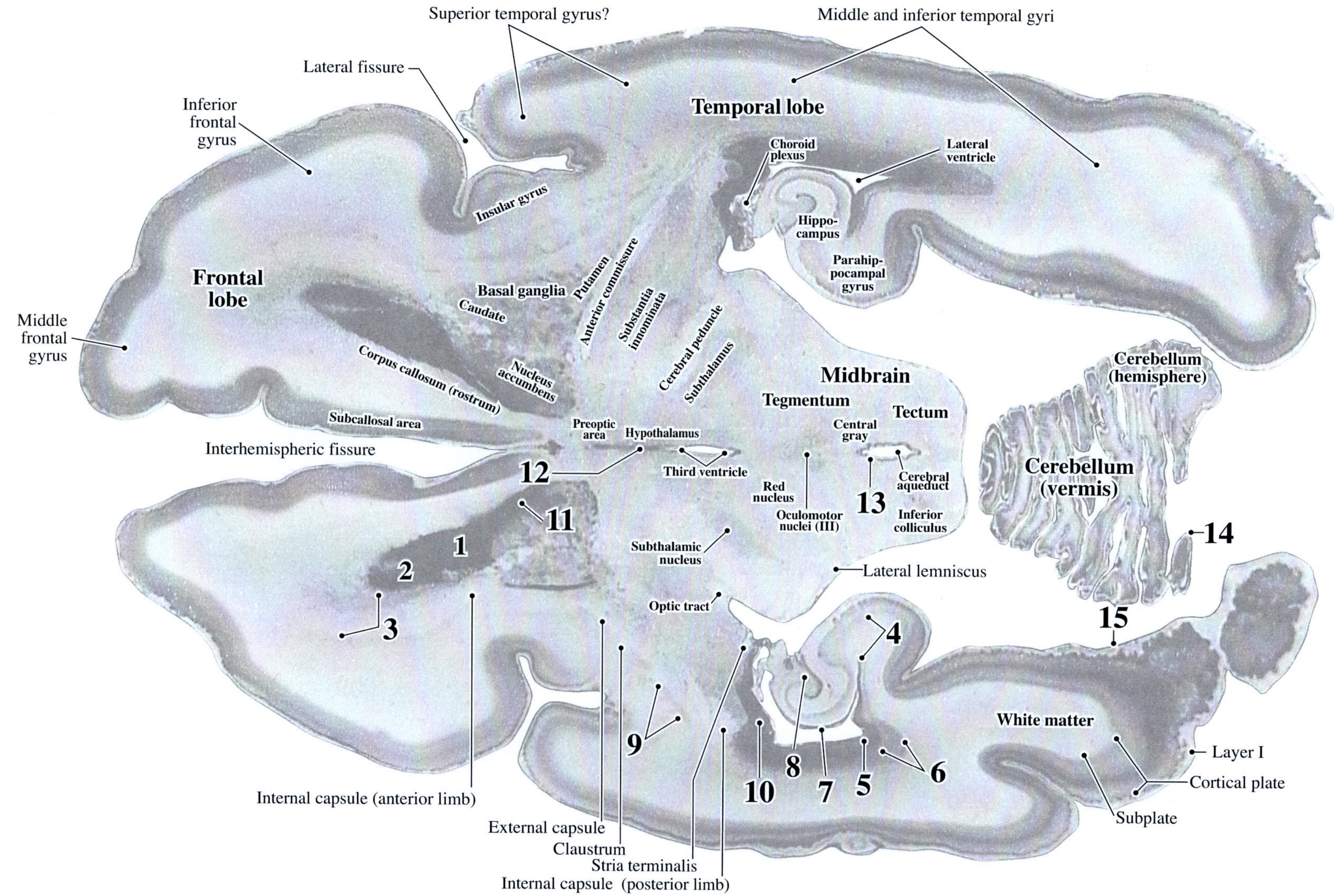

Superior temporal gyrus?
Middle and inferior temporal gyri
Lateral fissure
Inferior frontal gyrus
Temporal lobe
Choroid plexus
Lateral ventricle
Insular gyrus
Hippo-campus
Parahip-pocampal gyrus
Frontal lobe
Basal ganglia
Putamen
Anterior commissure
Caudate
Substantia innominata
Cerebral peduncle
Middle frontal gyrus
Corpus callosum (rostrum)
Nucleus accumbens
Subthalamus
Midbrain
Cerebellum (hemisphere)
Tegmentum
Tectum
Subcallosal area
Preoptic area
Hypothalamus
Central gray
Interhemispheric fissure
12
Third ventricle
Cerebral aqueduct
Cerebellum (vermis)
Red nucleus
13
Oculomotor nuclei (III)
Inferior colliculus
11
1
14
Subthalamic nucleus
2
Lateral lemniscus
3
Optic tract
15
4
White matter
9
Layer I
8
10
7
5
6
Cortical plate
Subplate
Internal capsule (anterior limb)
External capsule
Claustrum
Stria terminalis
Internal capsule (posterior limb)

PLATE 34A
CR 235 mm
GW 26, Y16-59
Horizontal
Section 660

See detail of the brain core and cerebellum in Plates 49A and B.

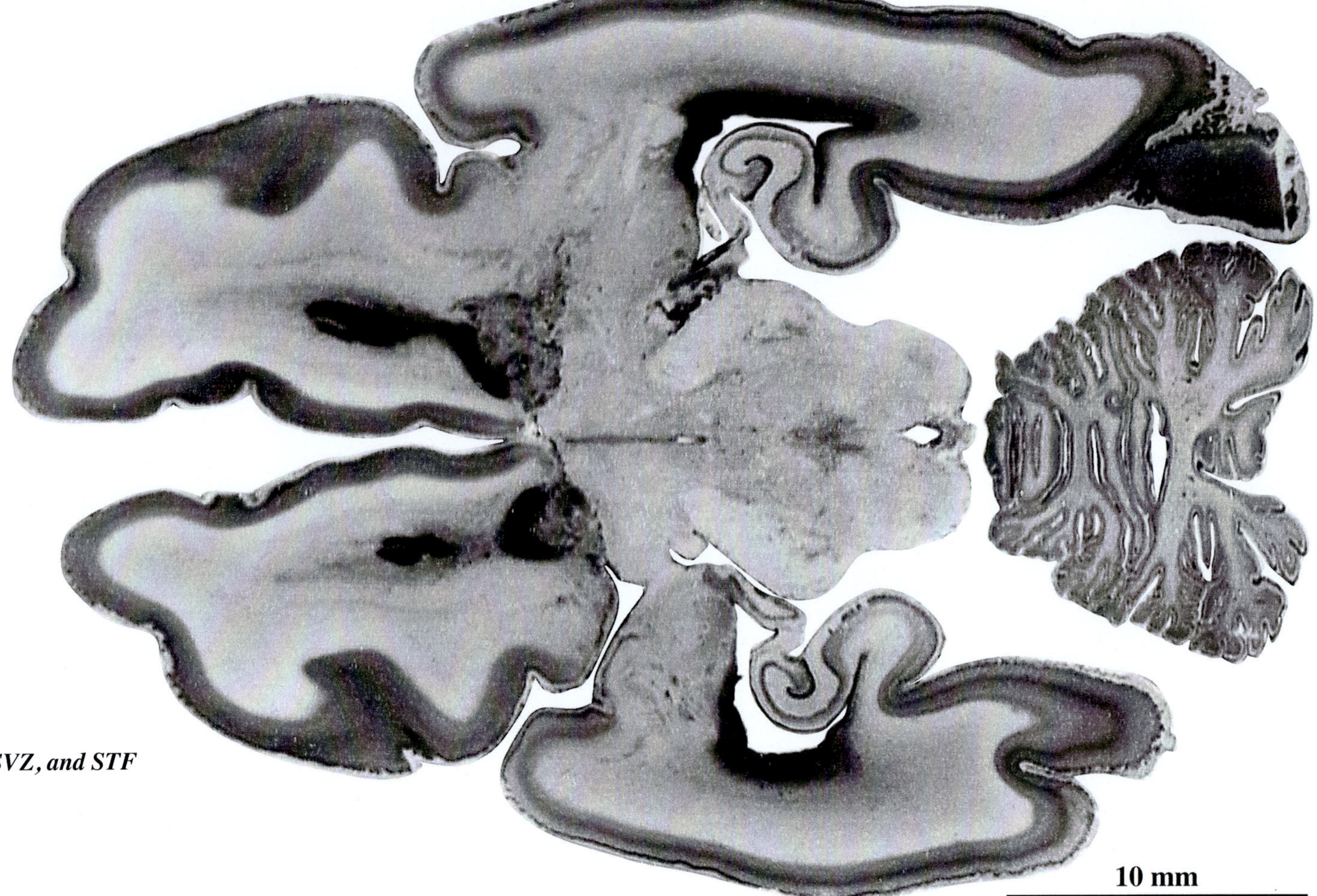

10 mm

Remnants of the germinal matrix, migratory streams, and transitional fields

1 *Rostral migratory stream*
2 *Frontal NEP and SVZ*
3 *Frontal STF*
4 *Parahippocampal NEP, SVZ, and STF*
5 *Temporal NEP and SVZ*
6 *Temporal STF*
7 *Alvear GEP*
8 *Subgranular zone (dentate)*
9 *Lateral migratory stream (cortical)*
10 *Amygdaloid G/EP*
11 *Diencephalic (hypothalamic) G/EP*
12 *Mesencephalic G/EP*
13 *External germinal layer (cerebellar)*
14 *Subpial granular layer (cortical)*

GEP - Glioepithelium
G/EP - Glioepithelium/ependyma
NEP - Neuroepithelium
STF - Stratified transitional field
SVZ - Subventricular zone

PLATE 34B

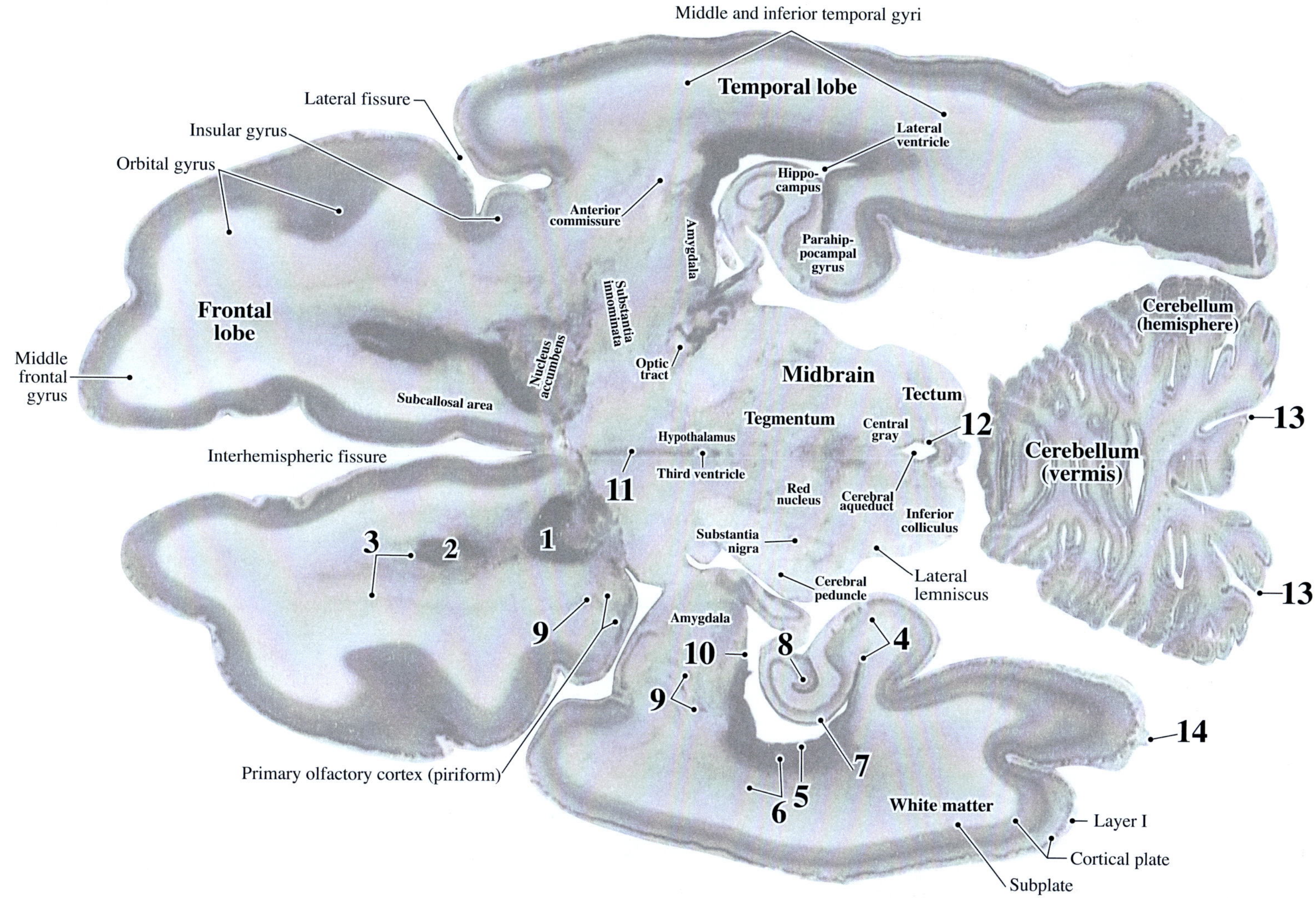

Middle and inferior temporal gyri
Temporal lobe
Lateral fissure
Insular gyrus
Orbital gyrus
Lateral ventricle
Hippo-campus
Anterior commissure
Amygdala
Parahip-pocampal gyrus
Substantia innominata
Frontal lobe
Cerebellum (hemisphere)
Nucleus accumbens
Optic tract
Middle frontal gyrus
Midbrain
Tectum
Subcallosal area
Tegmentum
Central gray
12
13
Hypothalamus
Cerebellum (vermis)
Interhemispheric fissure
Third ventricle
11
Red nucleus
Cerebral aqueduct
Inferior colliculus
1
3
2
Substantia nigra
Cerebral peduncle
Lateral lemniscus
13
9
Amygdala
4
10
8
9
14
7
Primary olfactory cortex (piriform)
5
6
White matter
Layer I
Cortical plate
Subplate

PLATE 35A
CR 235 mm
GW 26, Y16-59
Horizontal
Section 700

See detail of the brain core and cerebellum in Plates 50A and B.

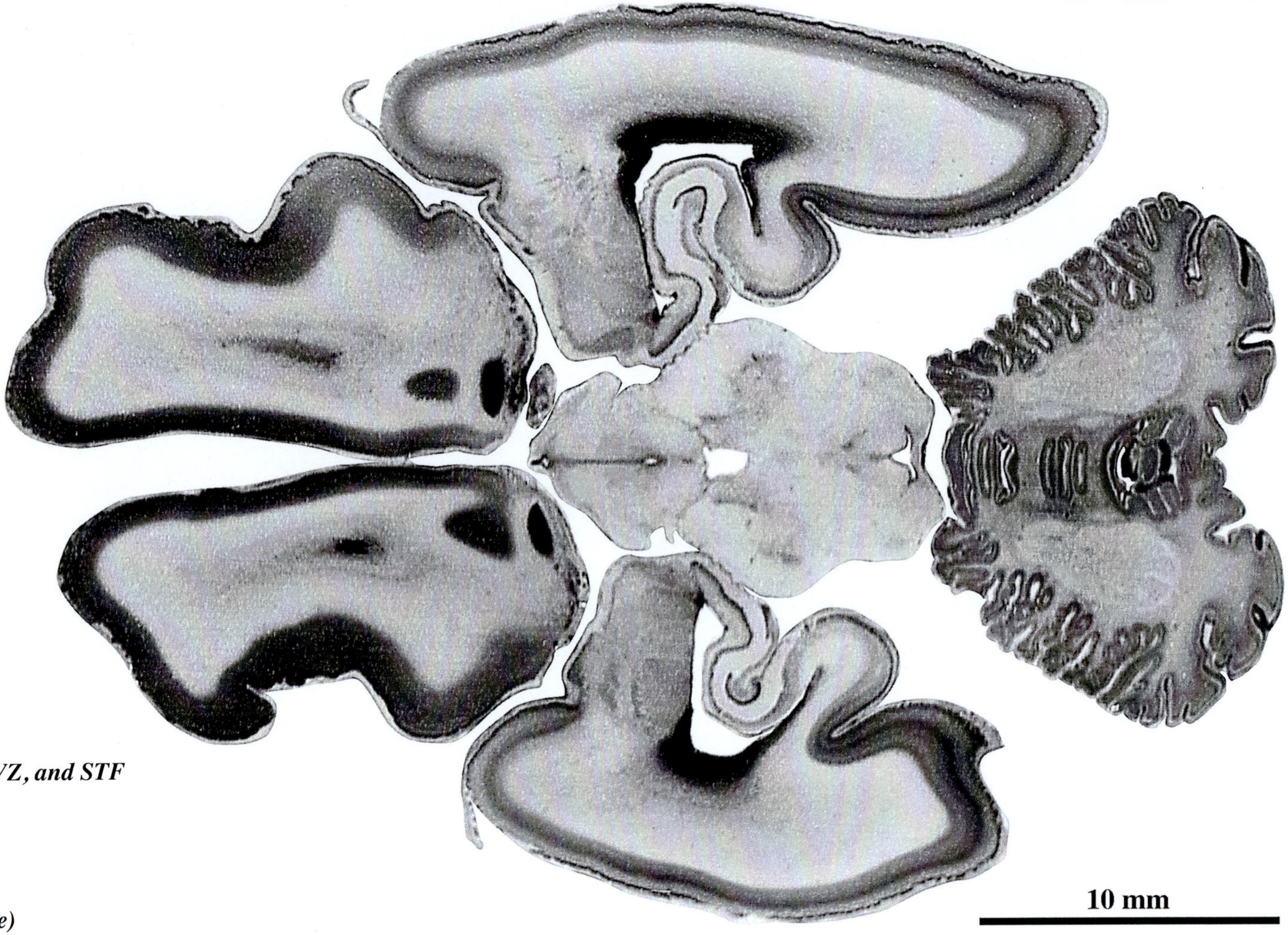

Remnants of the germinal matrix, migratory streams, and transitional fields

1 *Rostral migratory stream*
2 *Frontal NEP and SVZ*
3 *Frontal STF*
4 *Parahippocampal NEP, SVZ, and STF*
5 *Temporal NEP and SVZ*
6 *Temporal STF*
7 *Alvear GEP*
8 *Subgranular zone (dentate)*
9 *Lateral migratory stream (cortical)*
10 *Amygdaloid G/EP*
11 *Diencephalic (hypothalamic) G/EP*
12 *Mesencephalic G/EP*
13 *External germinal layer (cerebellar)*
14 *Subpial granular layer (cortical)*

GEP - Glioepithelium
G/EP - Glioepithelium/ependyma
NEP - Neuroepithelium
STF - Stratified transitional field
SVZ - Subventricular zone

PLATE 35B

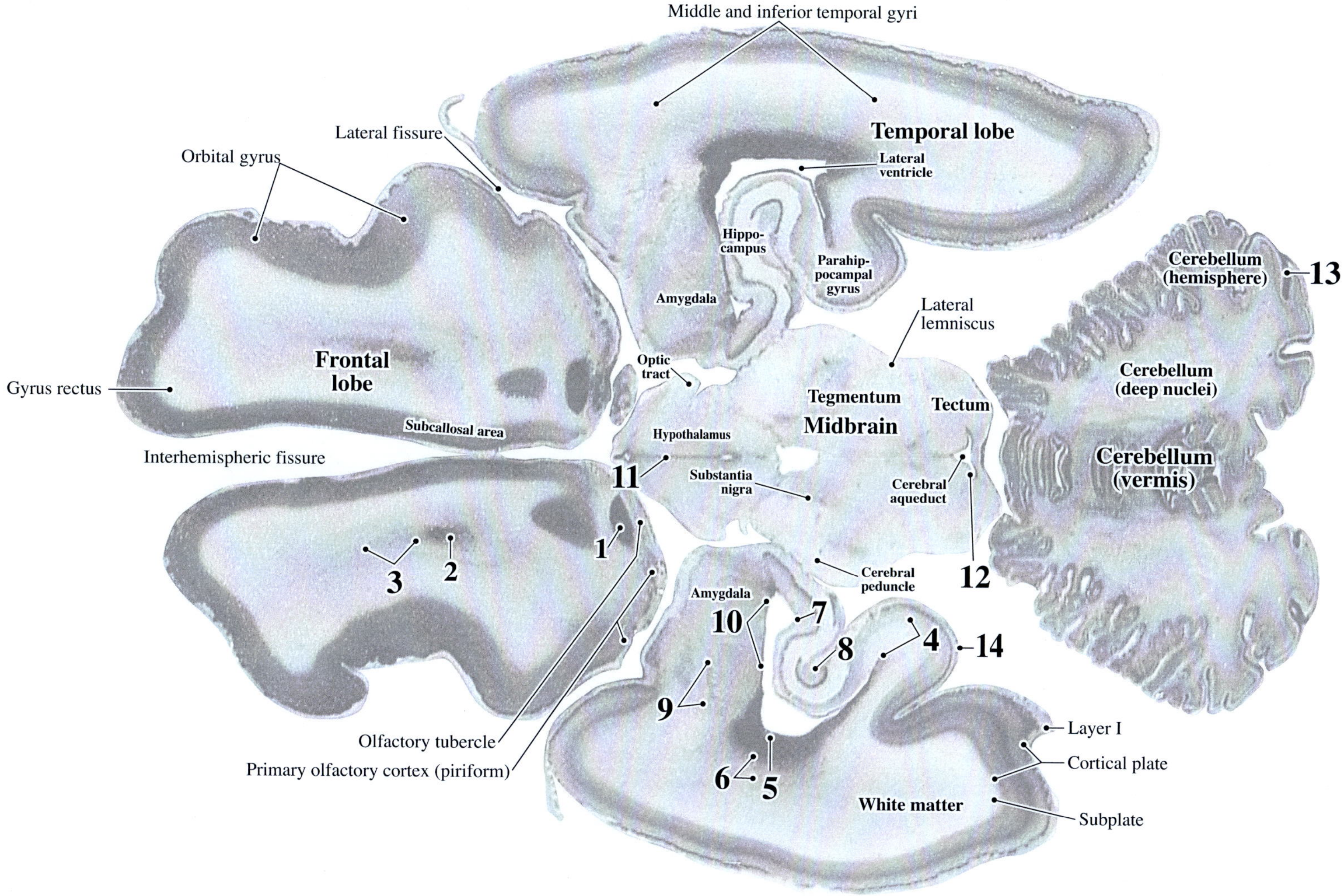

Middle and inferior temporal gyri
Temporal lobe
Lateral ventricle
Lateral fissure
Orbital gyrus
Hippo-campus
Parahip-pocampal gyrus
Amygdala
Cerebellum (hemisphere)
13
Lateral lemniscus
Frontal lobe
Optic tract
Cerebellum (deep nuclei)
Gyrus rectus
Tegmentum
Tectum
Midbrain
Subcallosal area
Hypothalamus
Interhemispheric fissure
Cerebellum (vermis)
11
Substantia nigra
Cerebral aqueduct
1
3
2
Cerebral peduncle
12
Amygdala
10
7
4
14
8
9
Layer I
Olfactory tubercle
Cortical plate
6
5
Primary olfactory cortex (piriform)
White matter
Subplate

PLATE 36A
CR 235 mm
GW 26, Y16-59
Horizontal
Section 740

See detail of the brain core and cerebellum in Plates 51A and B.

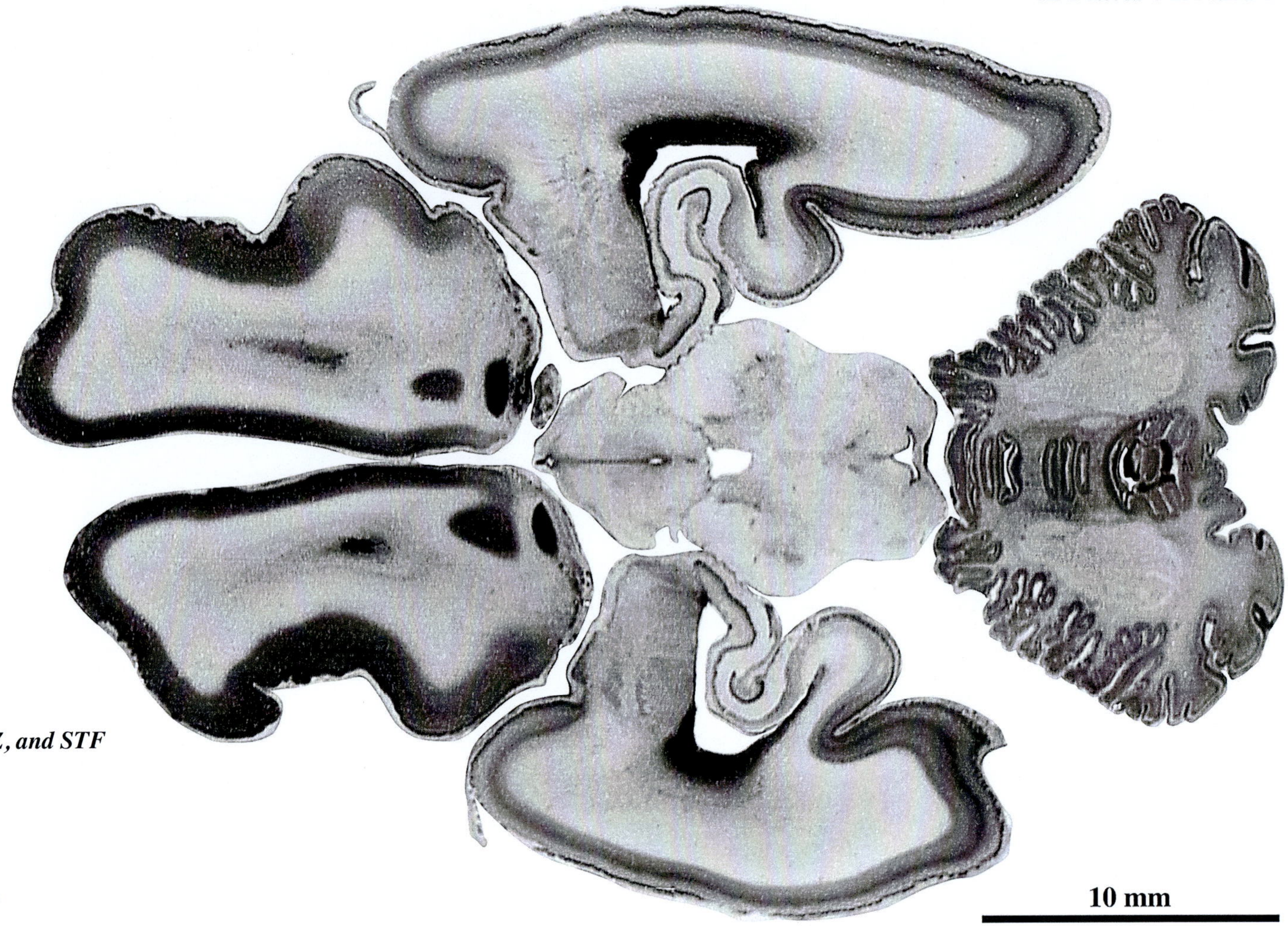

10 mm

Remnants of the germinal matrix, migratory streams, and transitional fields

1 *Rostral migratory stream*
2 *Frontal STF*
3 *Parahippocampal NEP, SVZ, and STF*
4 *Temporal NEP and SVZ*
5 *Temporal STF*
6 *Alvear GEP*
7 *Subgranular zone (dentate)*
8 *Lateral migratory stream (cortical)*
9 *Amygdaloid G/EP*
10 *Diencephalic (hypothalamic) G/EP*
11 *Pontine G/EP*
12 *External germinal layer (cerebellar)*
13 *Subpial granular layer (cortical)*

GEP - Glioepithelium
G/EP - Glioepithelium/ependyma
NEP - Neuroepithelium
STF - Stratified transitional field
SVZ - Subventricular zone

PLATE 36B

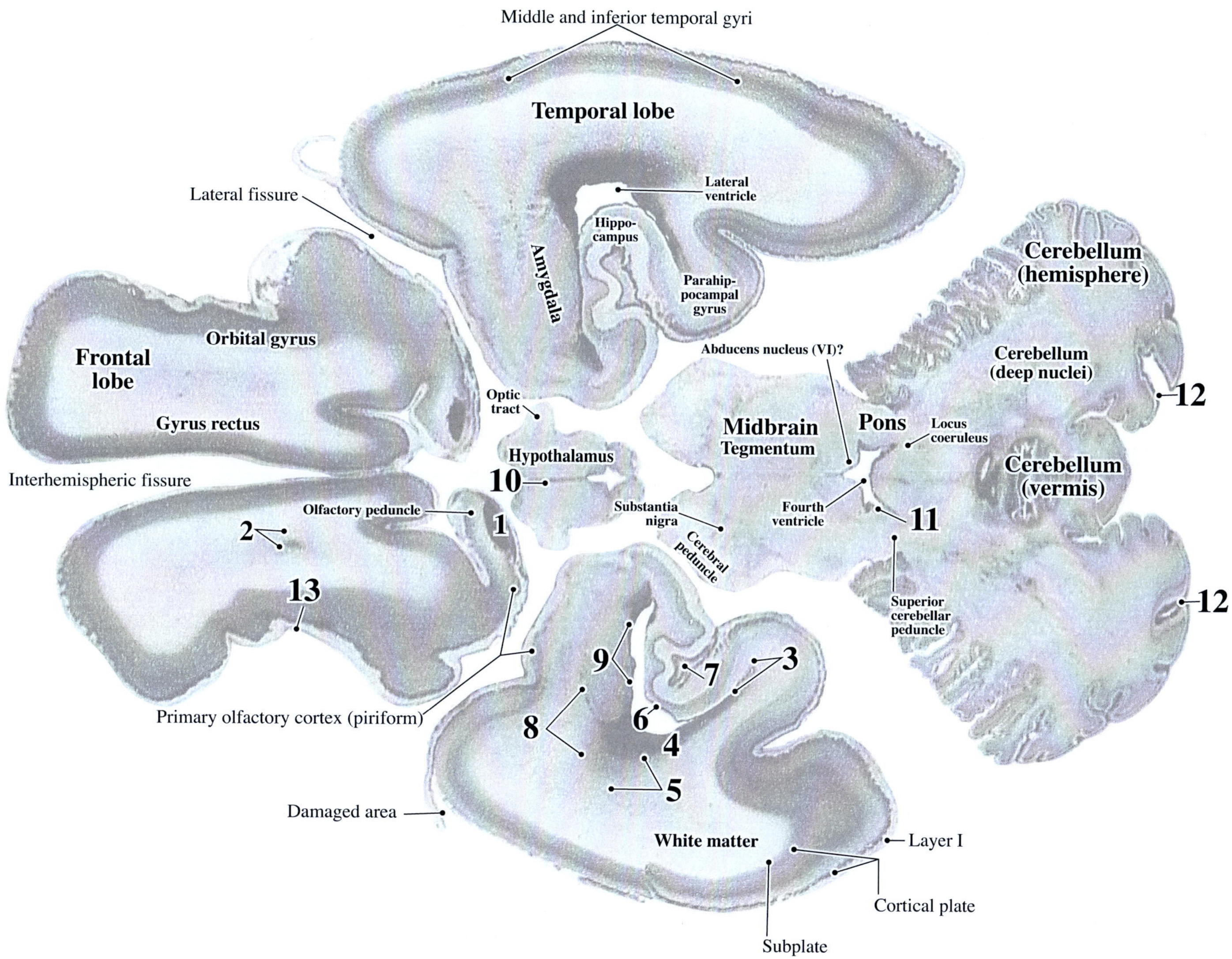

Middle and inferior temporal gyri
Temporal lobe
Lateral fissure
Lateral ventricle
Hippo-campus
Amygdala
Parahip-pocampal gyrus
Cerebellum (hemisphere)
Frontal lobe
Orbital gyrus
Abducens nucleus (VI)?
Cerebellum (deep nuclei)
12
Optic tract
Gyrus rectus
Midbrain Tegmentum
Pons
Locus coeruleus
Hypothalamus
Cerebellum (vermis)
Interhemispheric fissure
10
Substantia nigra
Fourth ventricle
11
Olfactory peduncle
1
2
Cerebral peduncle
13
Superior cerebellar peduncle
12
9
7
3
Primary olfactory cortex (piriform)
8
6
4
5
Damaged area
White matter
Layer I
Cortical plate
Subplate

PLATE 37A
CR 235 mm
GW 26, Y16-59
Horizontal
Section 780

See detail of the brain core and cerebellum in Plates 52A and B.

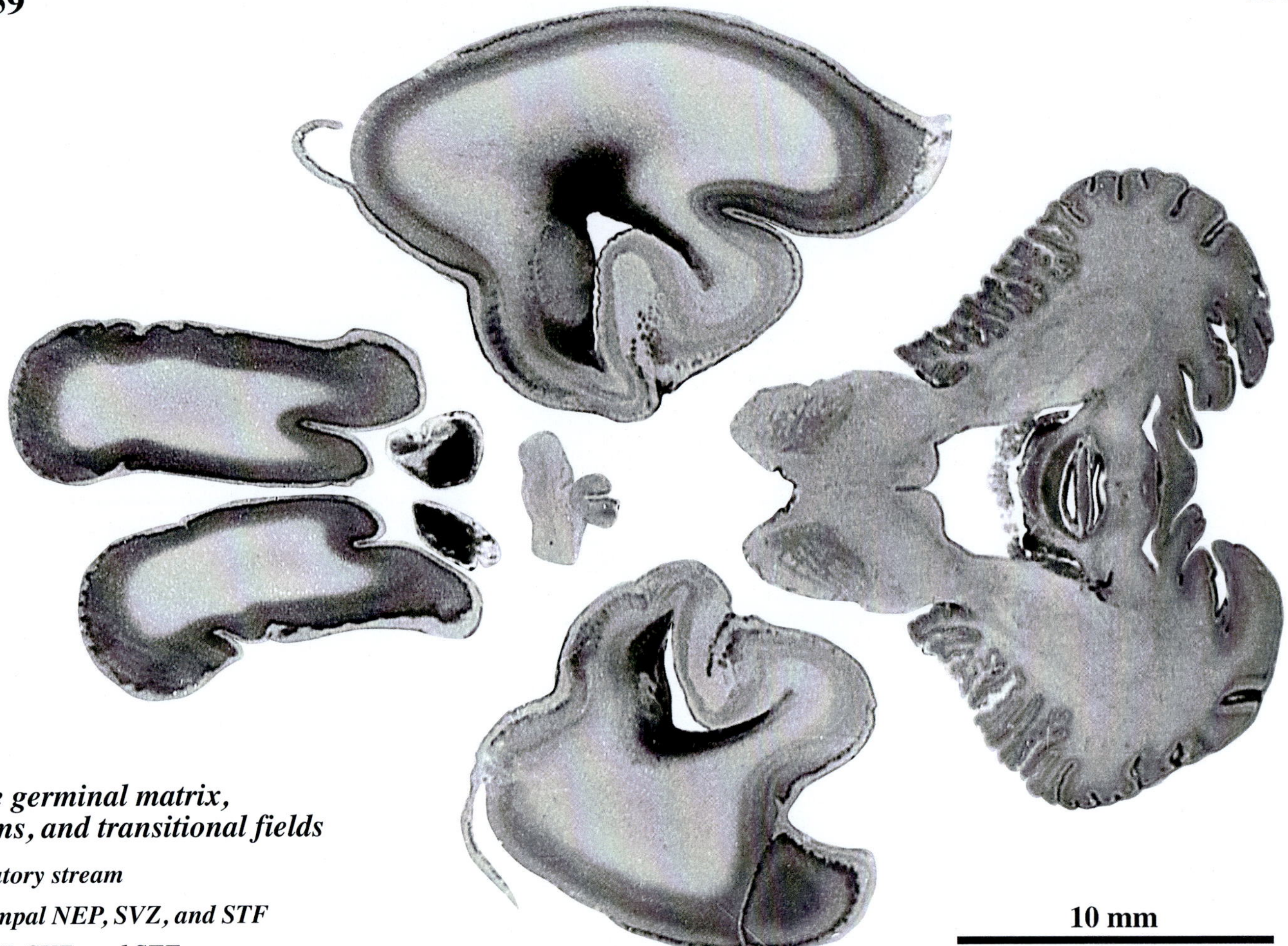

Remnants of the germinal matrix, migratory streams, and transitional fields

1 *Rostral migratory stream*
2 *Parahippocampal NEP, SVZ, and STF*
3 *Temporal NEP, SVZ, and STF*
4 *Alvear GEP*
5 *Lateral migratory stream (cortical)*
6 *Amygdaloid G/EP*
7 *Pontine G/EP*
8 *External germinal layer (cerebellar)*
9 *Subpial granular layer (cortical)*

GEP - Glioepithelium
G/EP - Glioepithelium/ependyma
NEP - Neuroepithelium
STF - Stratified transitional field
SVZ - Subventricular zone

PLATE 37B

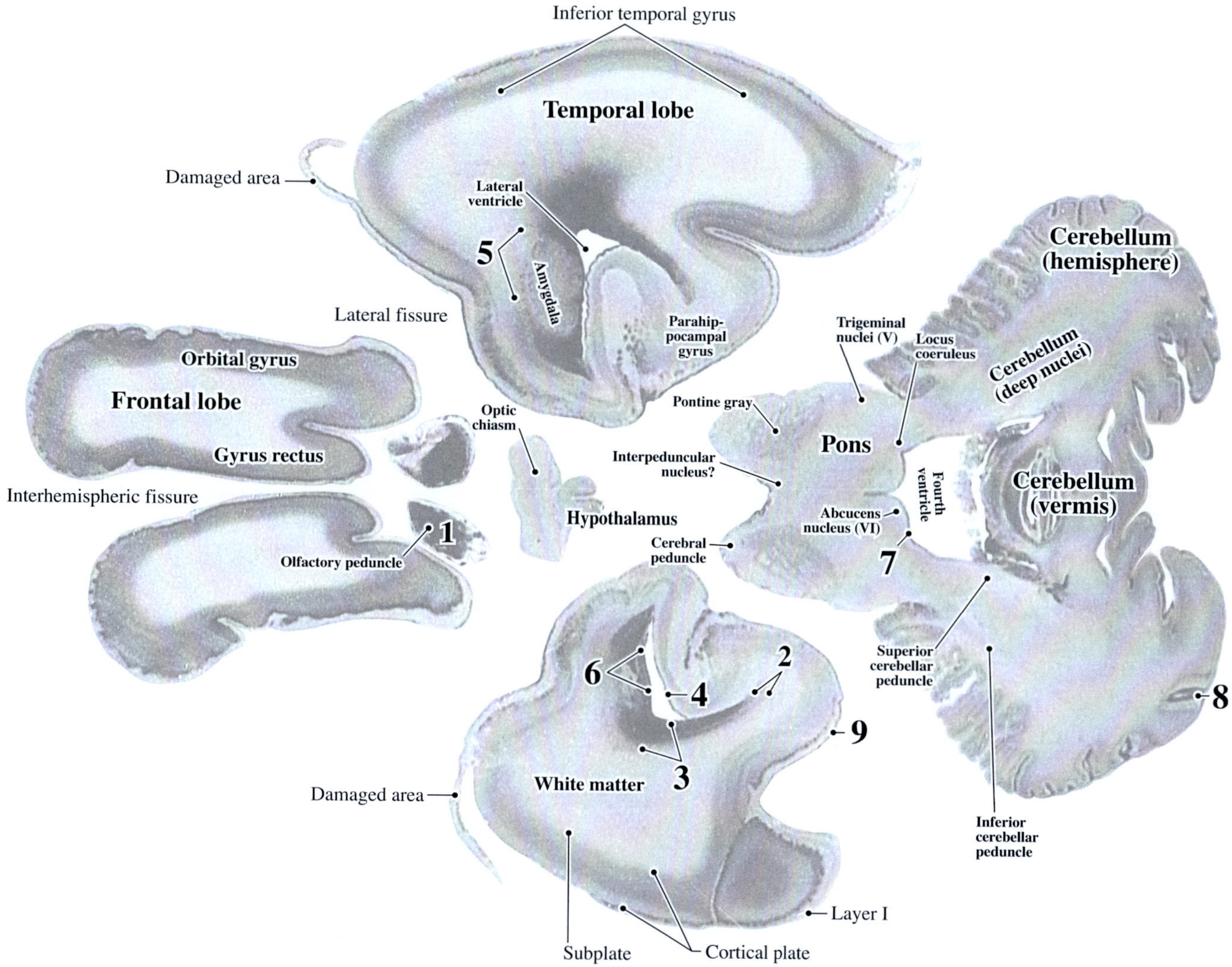
Inferior temporal gyrus
Temporal lobe
Damaged area
Lateral ventricle
5
Amygdala
Lateral fissure
Parahippocampal gyrus
Orbital gyrus
Frontal lobe
Gyrus rectus
Interhemispheric fissure
Optic chiasm
Olfactory peduncle
1
Hypothalamus
Pontine gray
Interpeduncular nucleus?
Trigeminal nuclei (V)
Locus coeruleus
Pons
Abcucens nucleus (VI)
Fourth ventricle
Cerebral peduncle
7
Cerebellum (hemisphere)
Cerebellum (deep nuclei)
Cerebellum (vermis)
6
4
2
9
3
White matter
Damaged area
Superior cerebellar peduncle
8
Inferior cerebellar peduncle
Layer I
Subplate
Cortical plate

PLATE 38A
CR 235 mm
GW 26, Y16-59
Horizontal
Section 820

See detail of the brain core and cerebellum in Plates 53A and B.

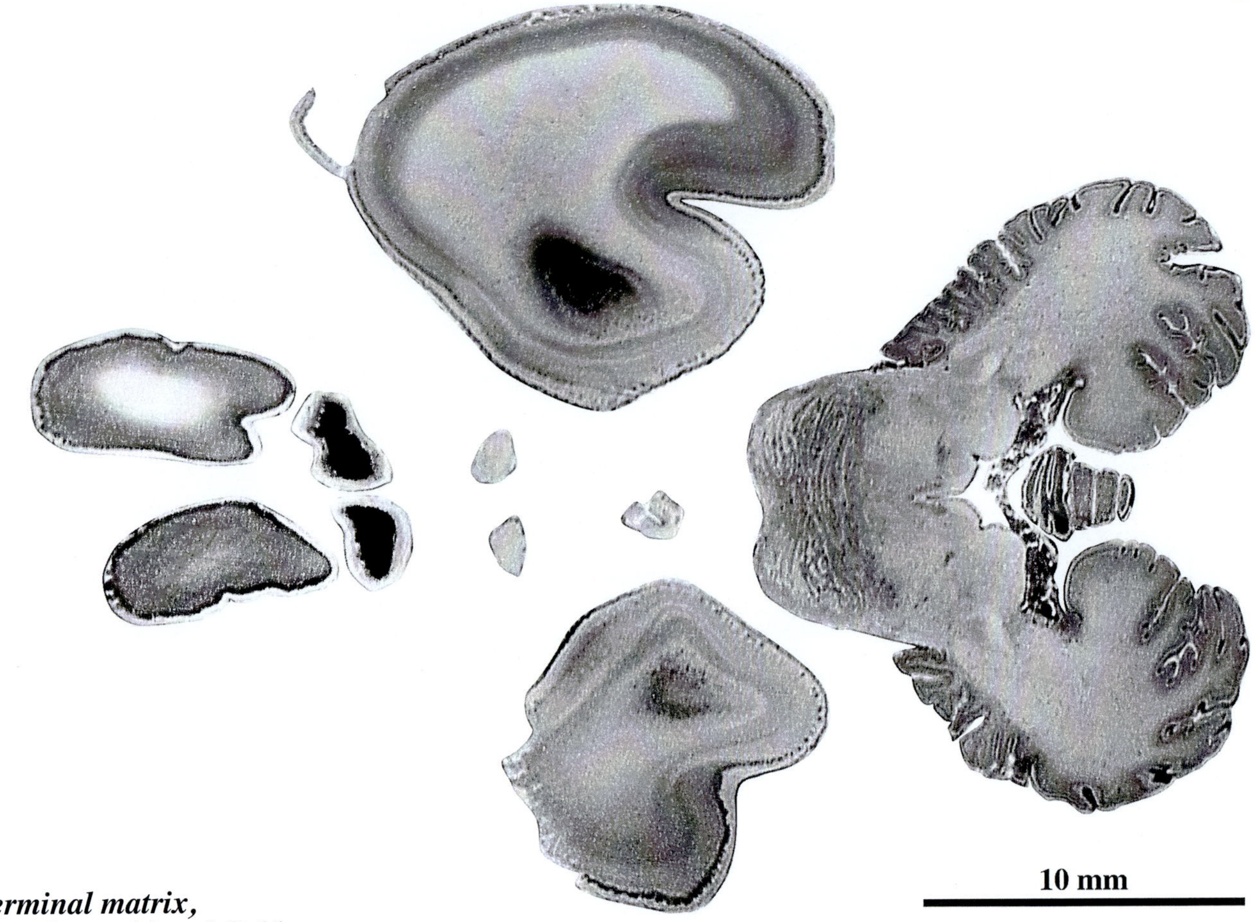

Remnants of the germinal matrix, migratory streams, and transitional fields

1 *Rostral migratory stream*
2 *Parahippocampal NEP, SVZ, and STF*
3 *Temporal NEP, SVZ, and STF*
4 *Amygdaloid G/EP*
5 *Medullary G/EP*
6 *External germinal layer (cerebellar)*
7 *Subpial granular layer (cortical)*

G/EP - Glioepithelium/ependyma
NEP - Neuroepithelium
STF - Stratified transitional field
SVZ - Subventricular zone

PLATE 38B

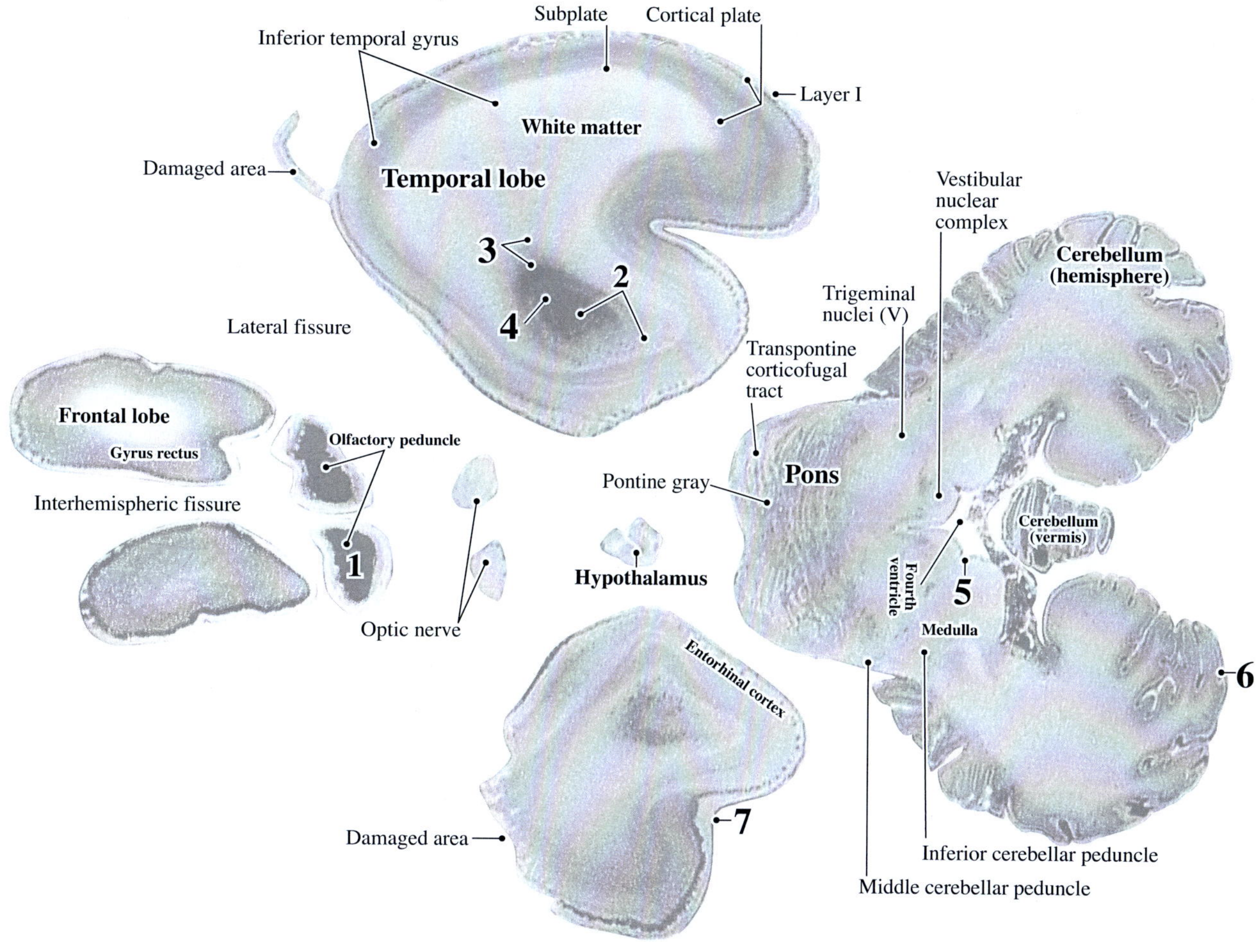
Subplate
Cortical plate
Inferior temporal gyrus
Layer I
White matter
Damaged area
Temporal lobe
Vestibular nuclear complex
3
2
4
Cerebellum (hemisphere)
Trigeminal nuclei (V)
Lateral fissure
Transpontine corticofugal tract
Frontal lobe
Olfactory peduncle
Gyrus rectus
Pontine gray
Pons
Interhemispheric fissure
Cerebellum (vermis)
1
Fourth ventricle
5
Hypothalamus
Medulla
Optic nerve
Entorhinal cortex
6
7
Damaged area
Inferior cerebellar peduncle
Middle cerebellar peduncle

PLATE 39A
CR 235 mm
GW 26, Y16-59
Horizontal
Section 860

See detail of the brain core and cerebellum in Plates 54A and B.

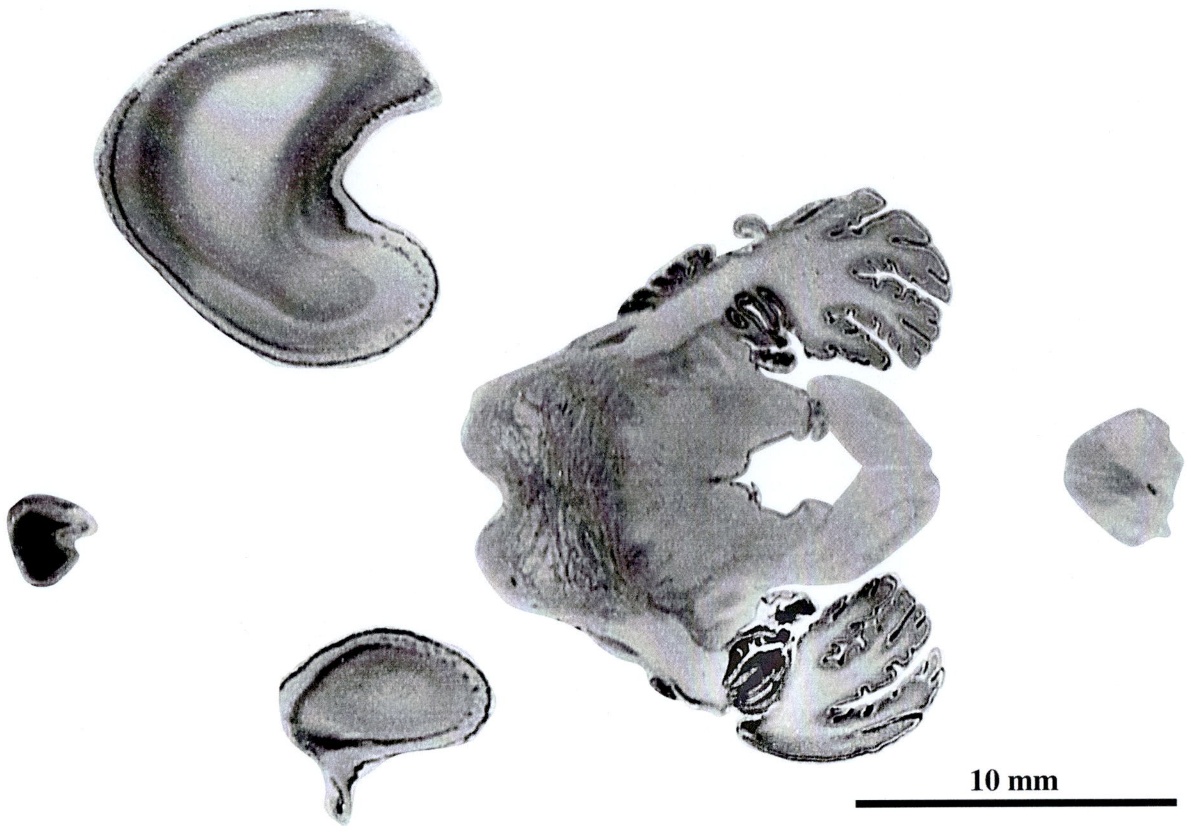

Remnants of the germinal matrix, migratory streams, and transitional fields

1 *Rostral migratory stream*

2 *Medullary glioepithelium/ependyma*

3 *External germinal layer (cerebellar)*

4 *Subpial granular layer (cortical)*

5 *Dorsal funicular myelination gliosis*

PLATE 39B

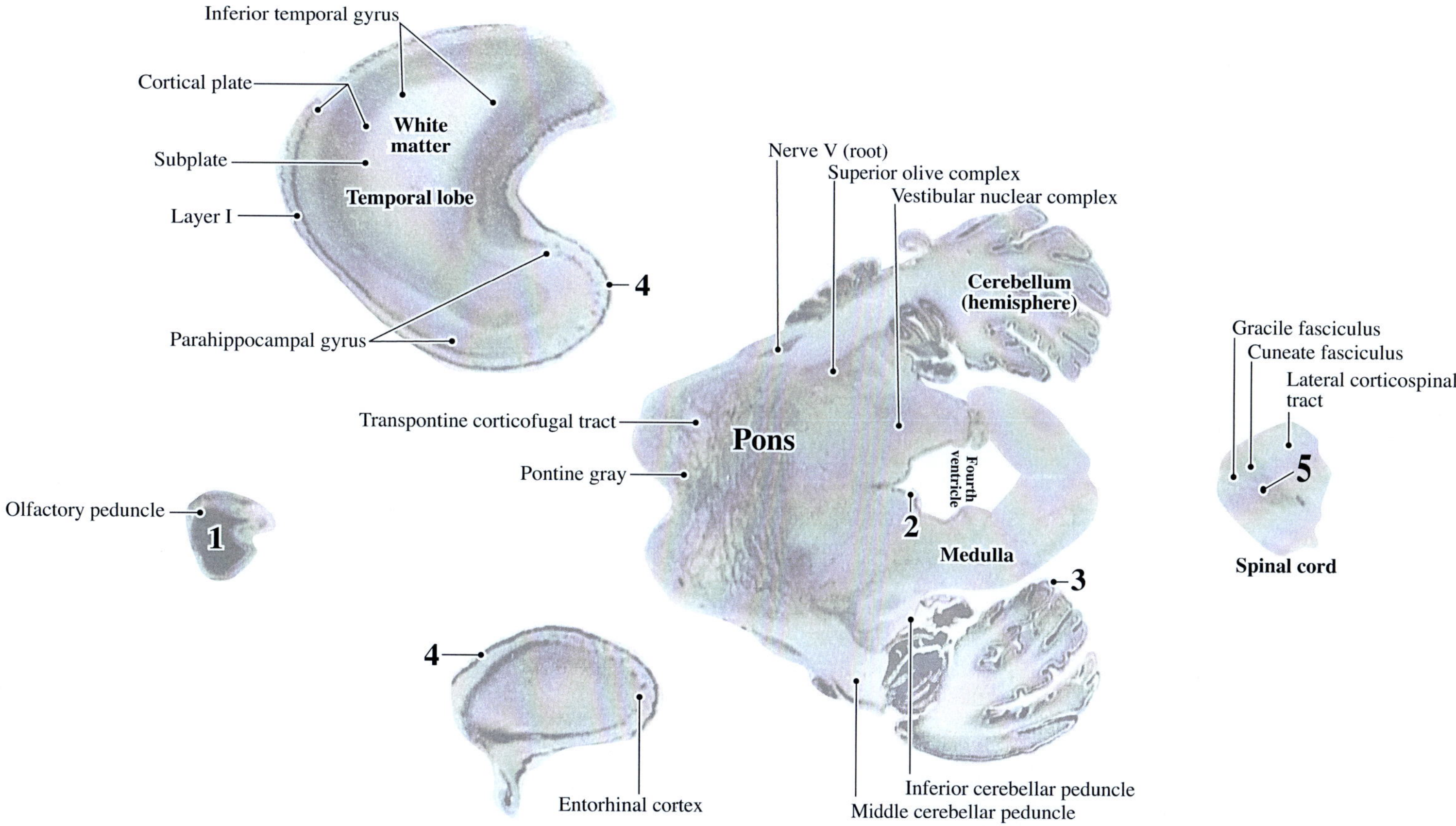
Inferior temporal gyrus
Cortical plate
White matter
Subplate
Temporal lobe
Layer I
4
Parahippocampal gyrus
Nerve V (root)
Superior olive complex
Vestibular nuclear complex
Cerebellum (hemisphere)
Transpontine corticofugal tract
Pons
Fourth ventricle
Pontine gray
Olfactory peduncle
1
2
Medulla
3
4
Entorhinal cortex
Inferior cerebellar peduncle
Middle cerebellar peduncle
Gracile fasciculus
Cuneate fasciculus
Lateral corticospinal tract
5
Spinal cord

PLATE 40A
CR 235 mm
GW 26, Y16-59
Horizontal
Section 880

See detail of the brain core and cerebellum in Plates 55A and B.

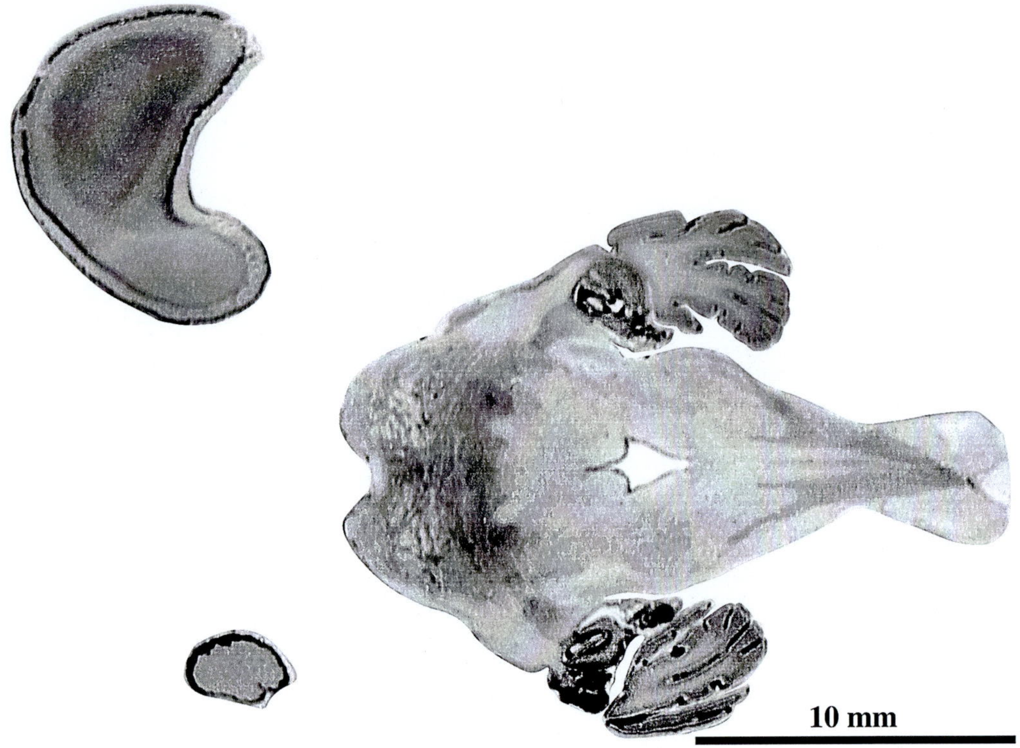

Remnants of the germinal matrix, migratory streams, and transitional fields

1 *Medullary glioepithelium/ependyma*

2 *External germinal layer (cerebellar)*

3 *Subpial granular layer (cortical)*

4 *Dorsal funicular myelination gliosis*

PLATE 40B

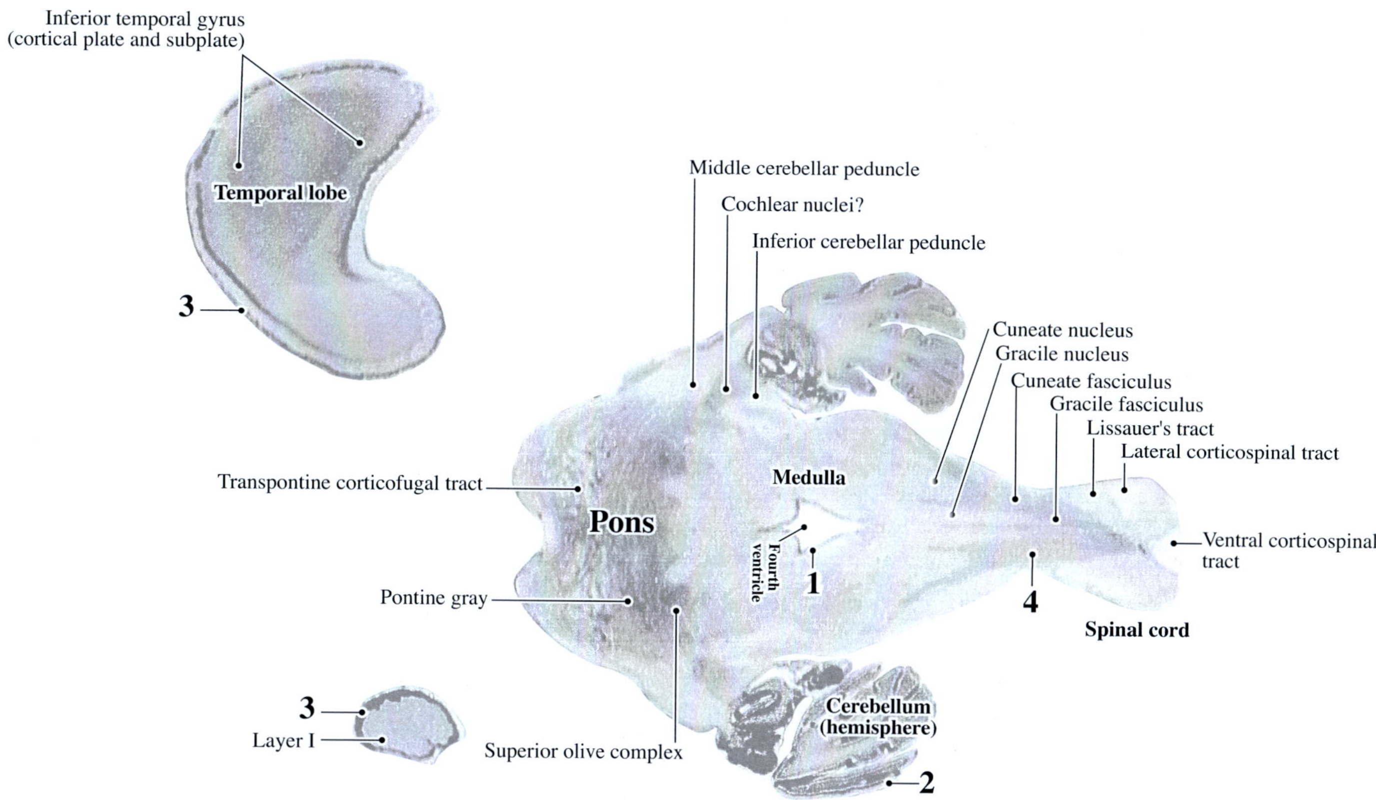
Inferior temporal gyrus
(cortical plate and subplate)
Temporal lobe
3
Middle cerebellar peduncle
Cochlear nuclei?
Inferior cerebellar peduncle
Cuneate nucleus
Gracile nucleus
Cuneate fasciculus
Gracile fasciculus
Lissauer's tract
Lateral corticospinal tract
Transpontine corticofugal tract
Medulla
Pons
Fourth ventricle
1
Ventral corticospinal tract
4
Spinal cord
Pontine gray
3
Layer I
Superior olive complex
Cerebellum (hemisphere)
2

PLATE 41A
CR 235 mm
GW 26, Y16-59
Horizontal
Section 900

See detail of the brain core
and cerebellum
in Plates 56A and B.

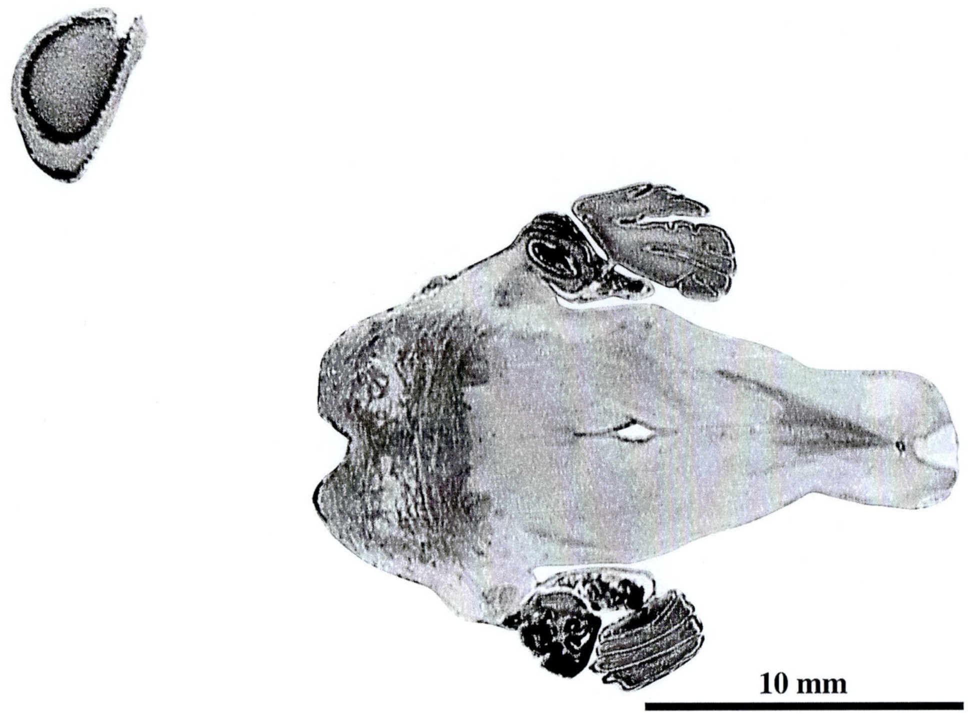

Remnants of the germinal matrix, migratory streams, and transitional fields

1 *Medullary glioepithelium/ependyma*

2 *External germinal layer (cerebellar)*

3 *Subpial granular layer (cortical)*

4 *Dorsal funicular myelination gliosis*

5 *Ventral funicular myelination gliosis*

PLATE 41B

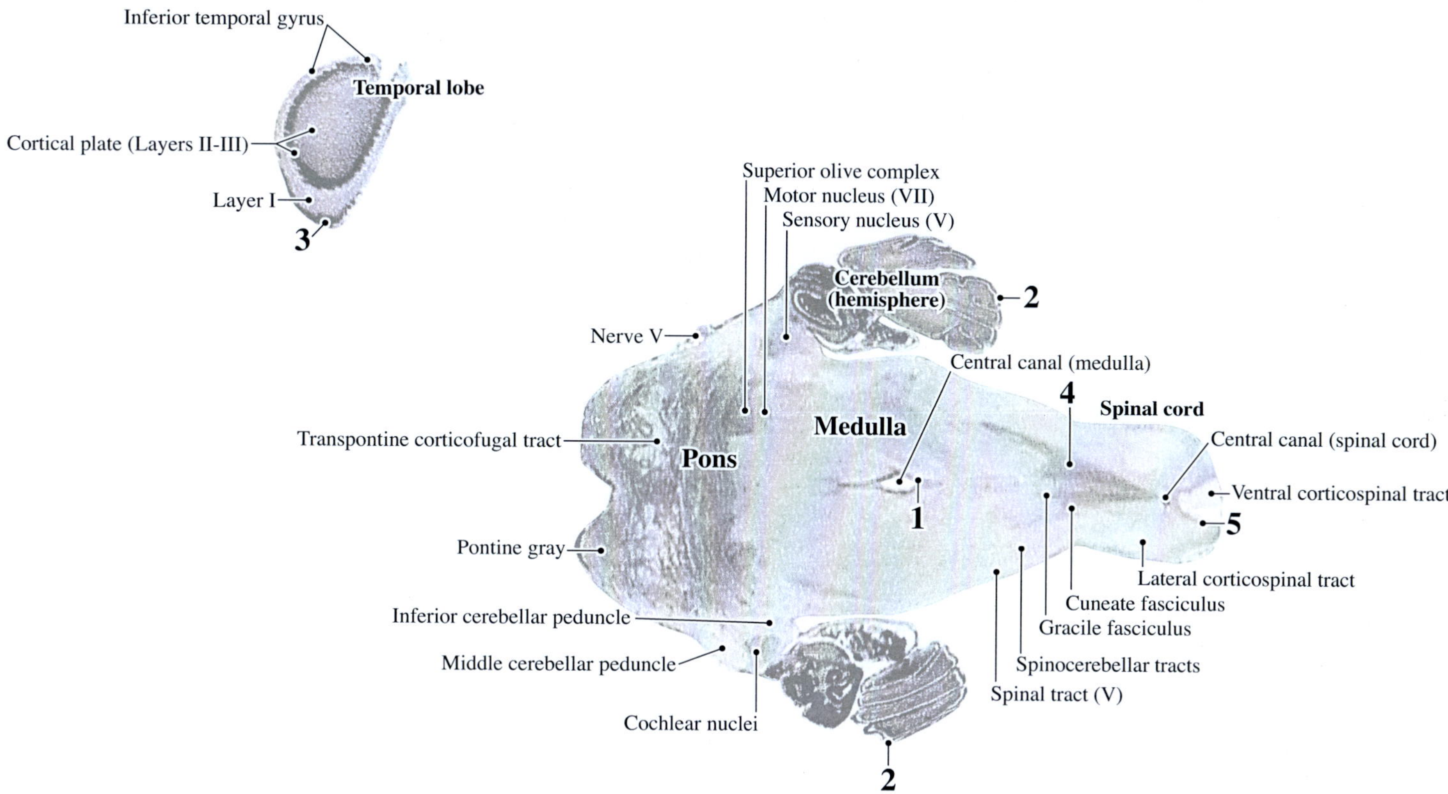

Inferior temporal gyrus
Temporal lobe
Cortical plate (Layers II-III)
Layer I
3
Superior olive complex
Motor nucleus (VII)
Sensory nucleus (V)
Cerebellum (hemisphere)
2
Nerve V
Central canal (medulla)
4
Spinal cord
Medulla
Transpontine corticofugal tract
Pons
Central canal (spinal cord)
Ventral corticospinal tract
1
5
Pontine gray
Lateral corticospinal tract
Cuneate fasciculus
Gracile fasciculus
Inferior cerebellar peduncle
Middle cerebellar peduncle
Spinocerebellar tracts
Spinal tract (V)
Cochlear nuclei
2

PLATE 42A
CR 235 mm
GW 26, Y16-59
Horizontal
Section 940

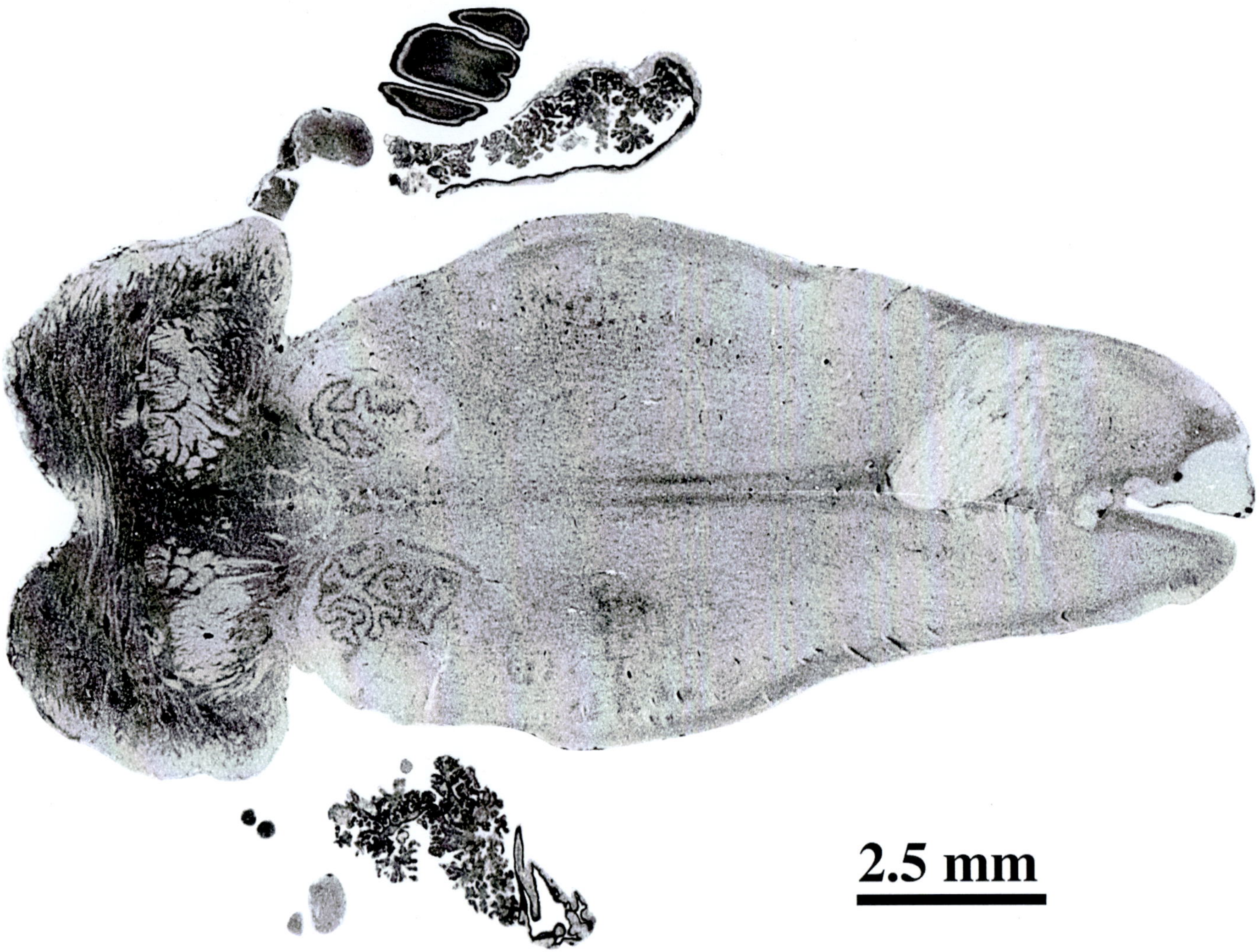

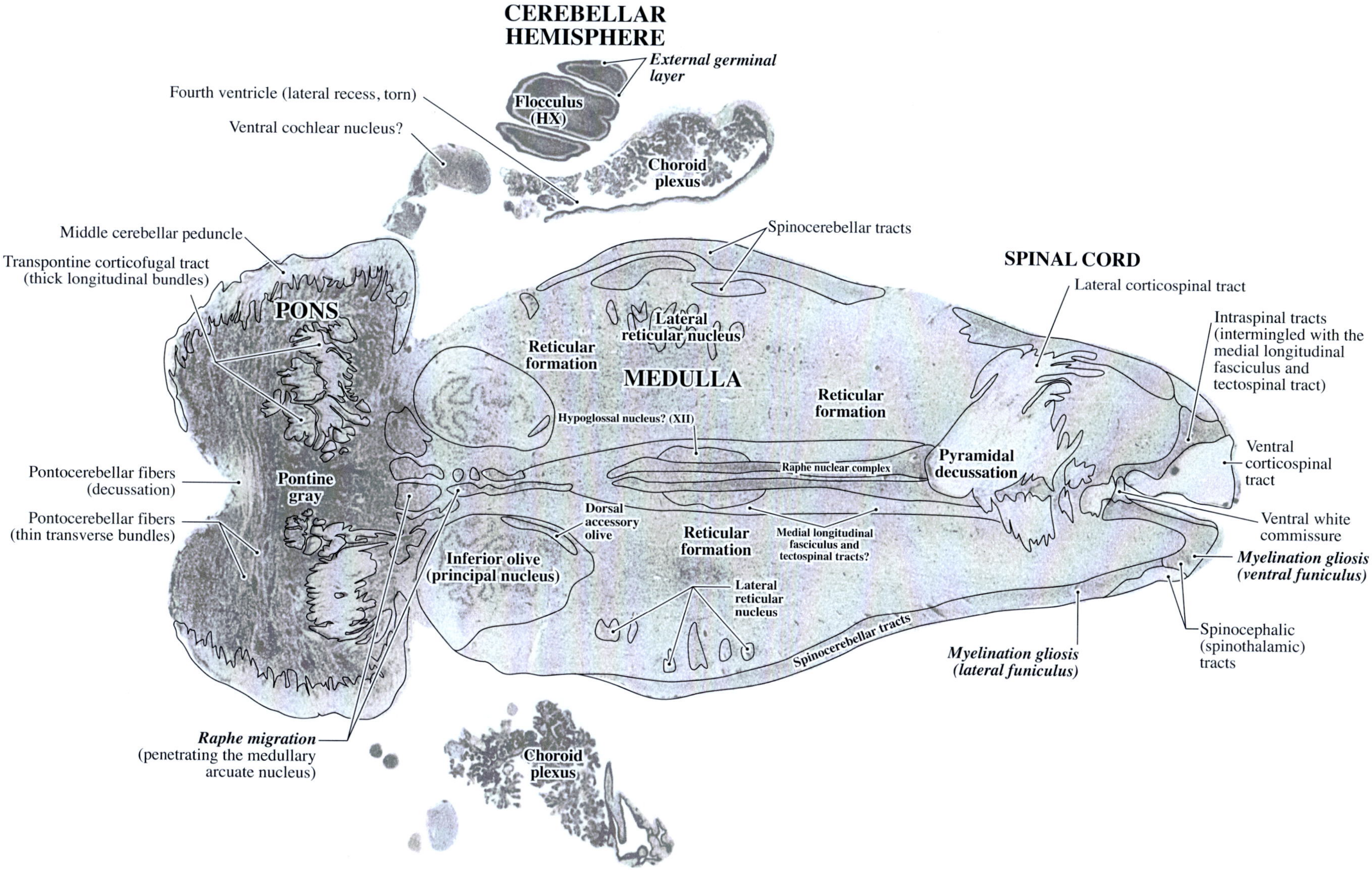

Germinal and transitional structures in *italics*

PLATE 43A
CR 235 mm
GW 26, Y16-59
Horizontal
Section 960

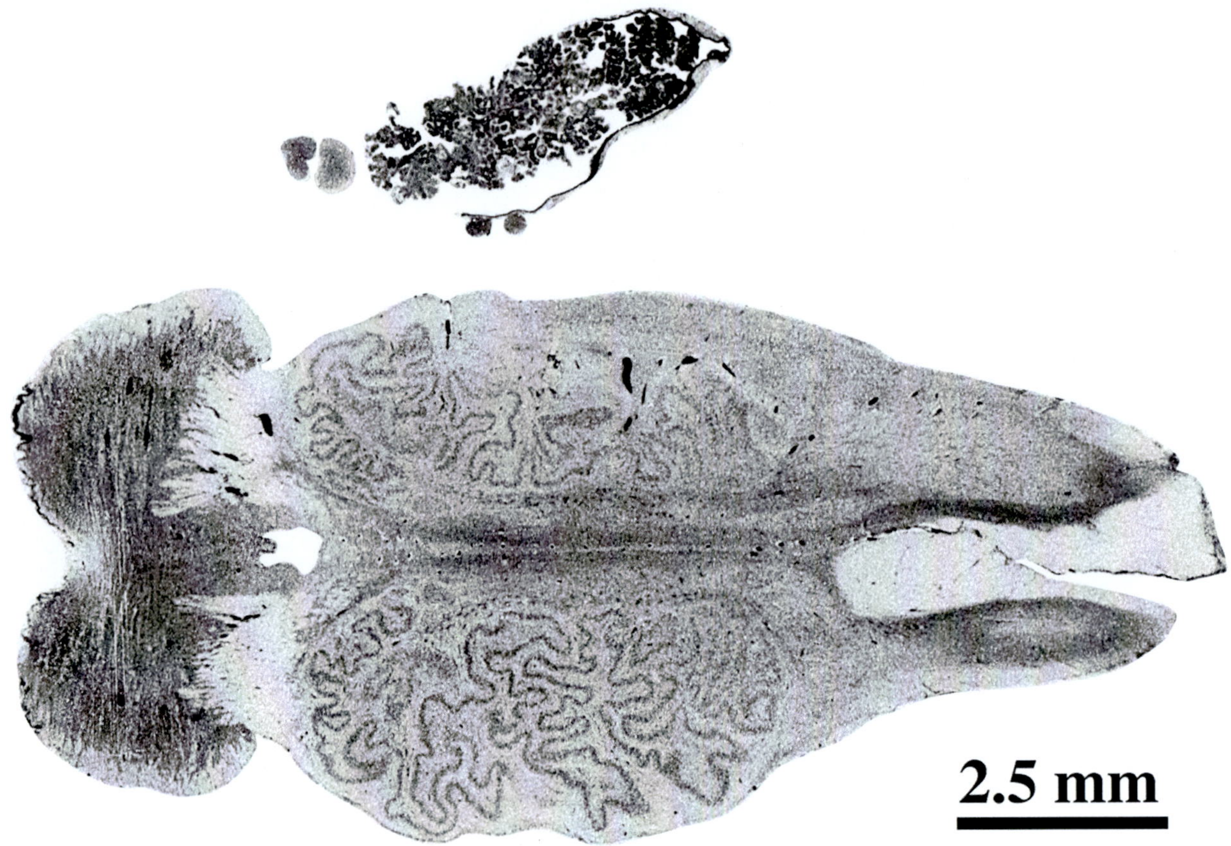

PLATE 43B

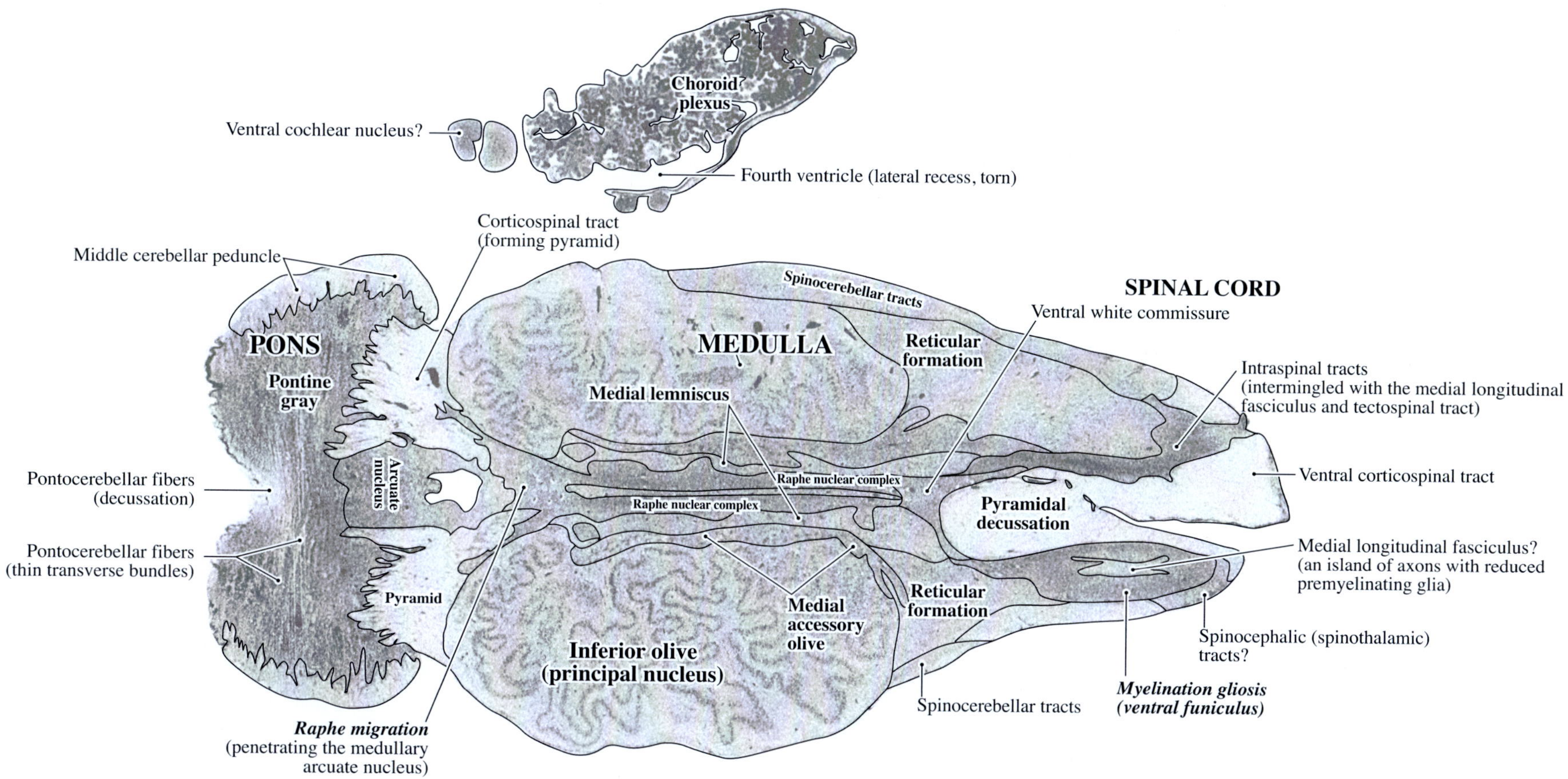

Germinal and transitional structures in *italics*

PLATE 44A, CR 235 mm, GW 26, Y16-59
Horizontal, Section 400

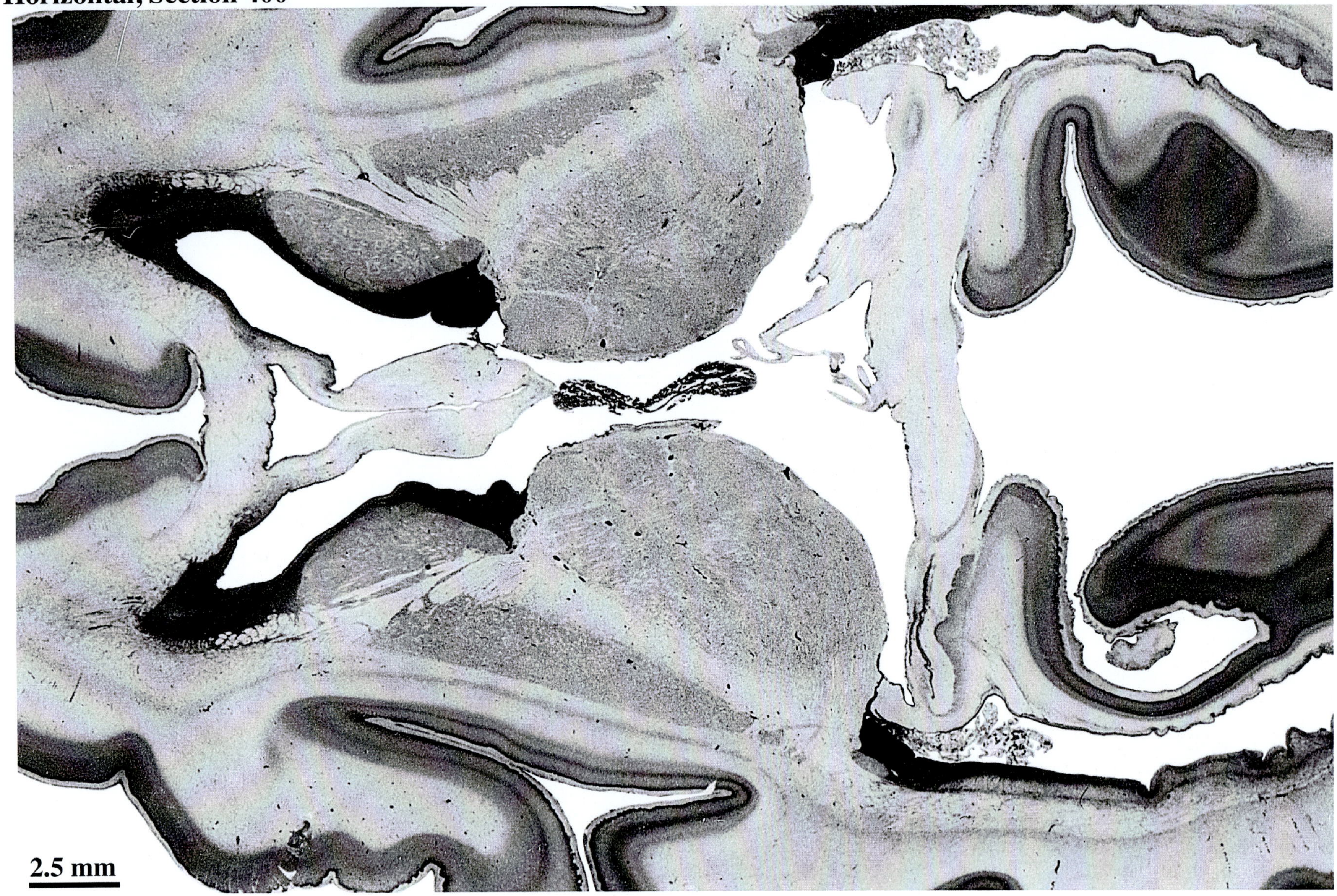

See the entire Section 400 in Plate 29.

PLATE 44B

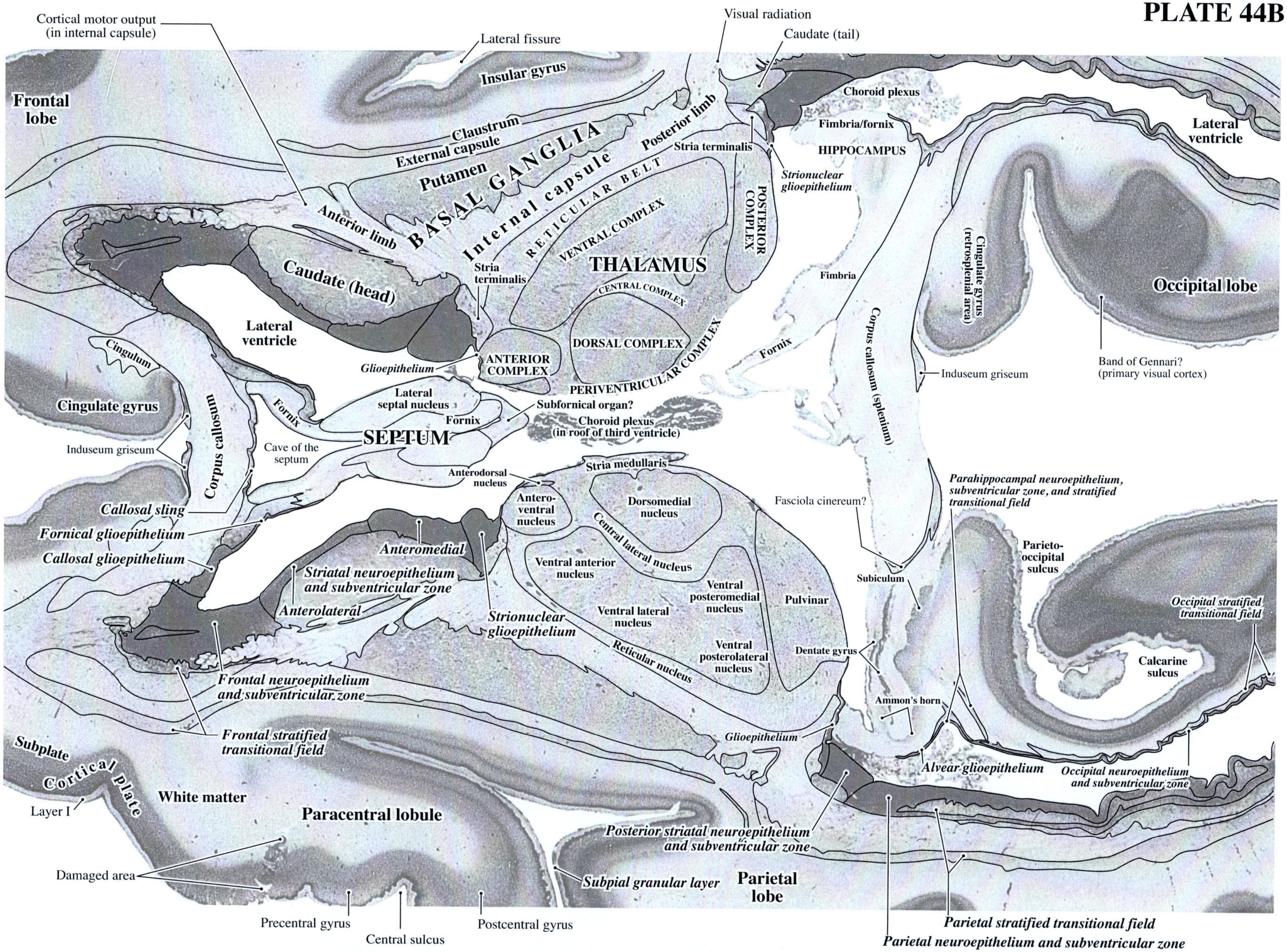

Germinal and transitional structures in *italics*

PLATE 45A
CR 235 mm
GW 26, Y16-59
Horizontal
Section 460

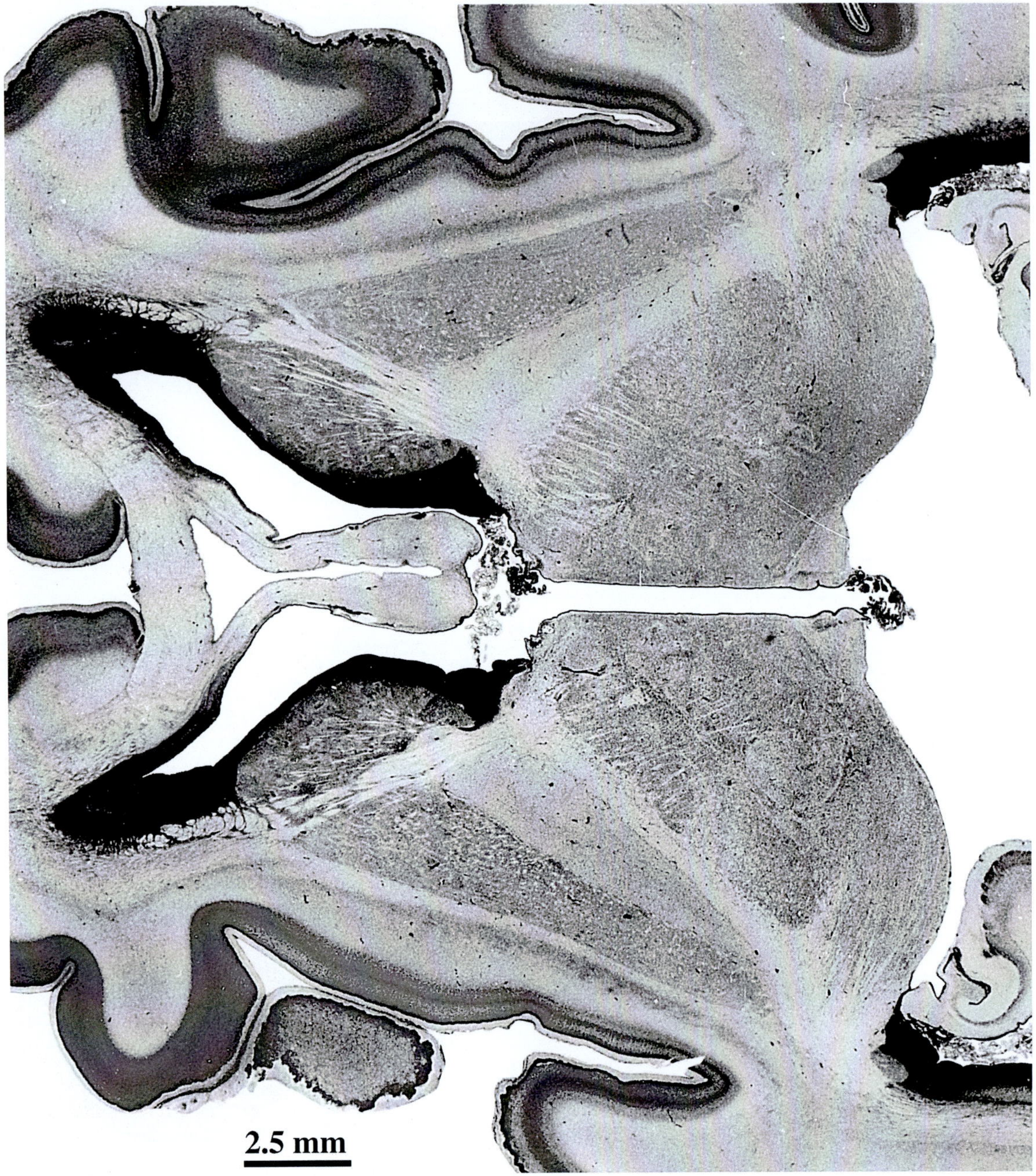

2.5 mm

See the entire Section 460 in Plate 30.

PLATE 45B

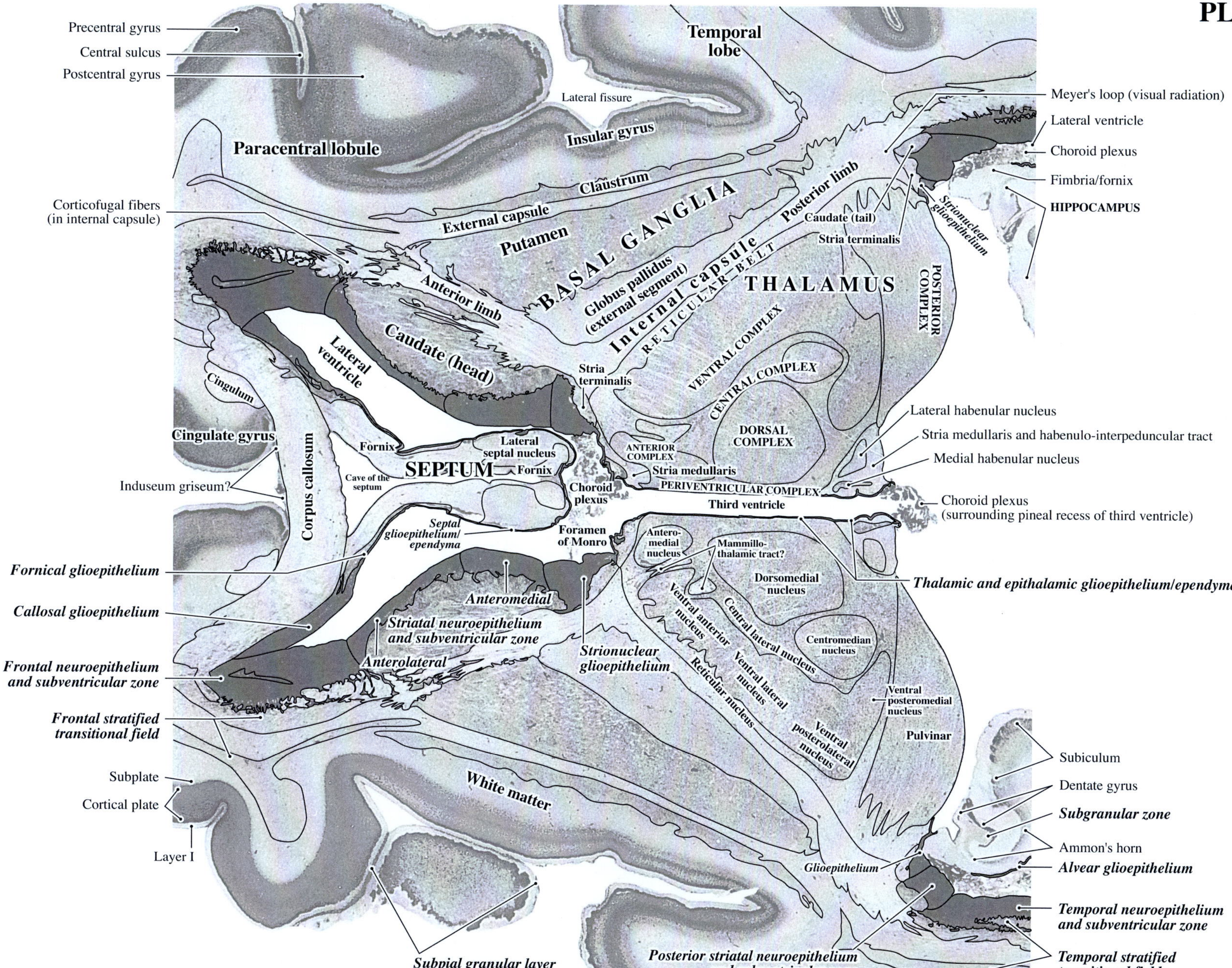

Germinal and transitional structures in *italics*

PLATE 46A
CR 235 mm
GW 26, Y16-59
Horizontal
Section 520

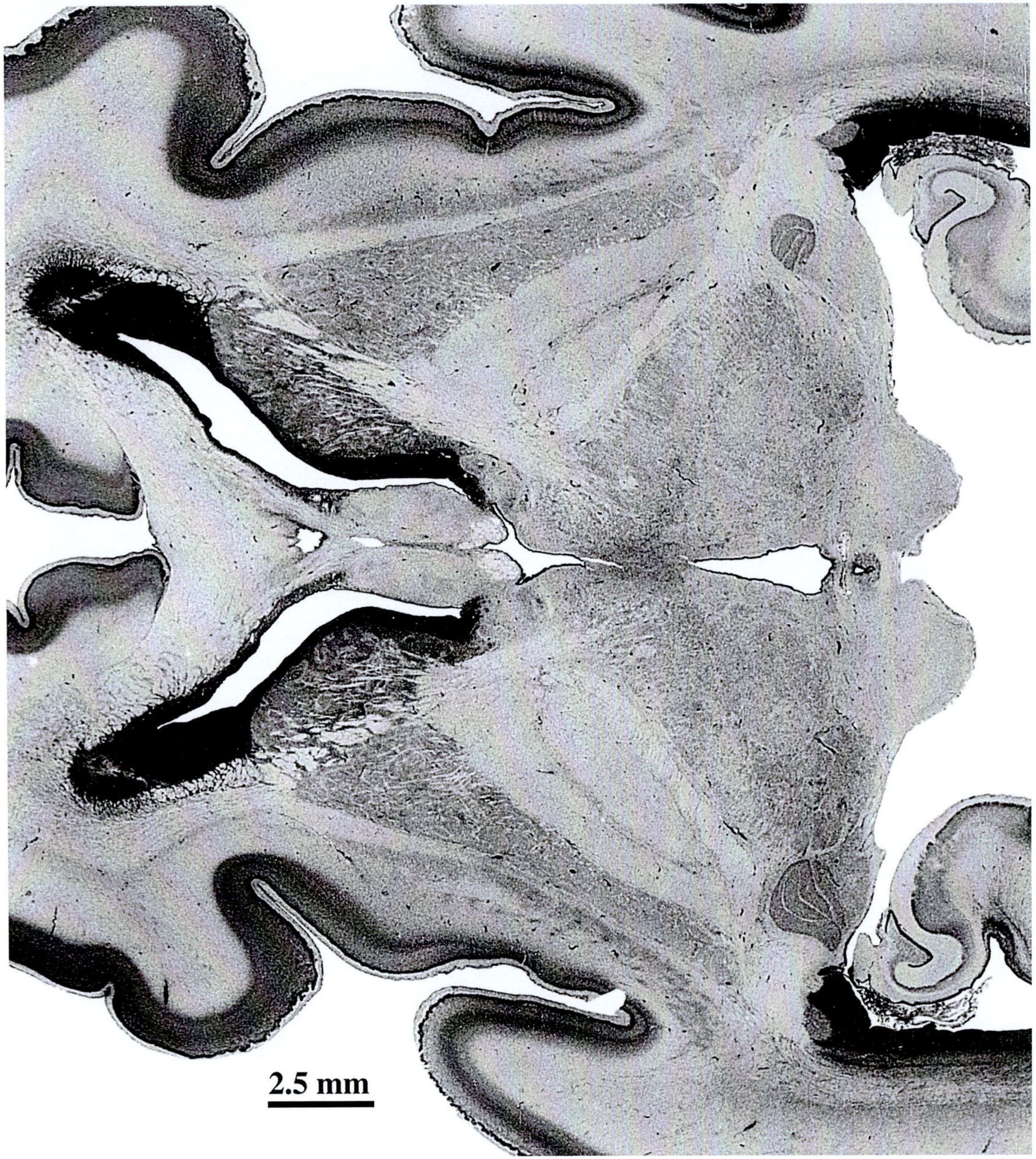

See the entire Section 520 in Plate 31.

PLATE 46B

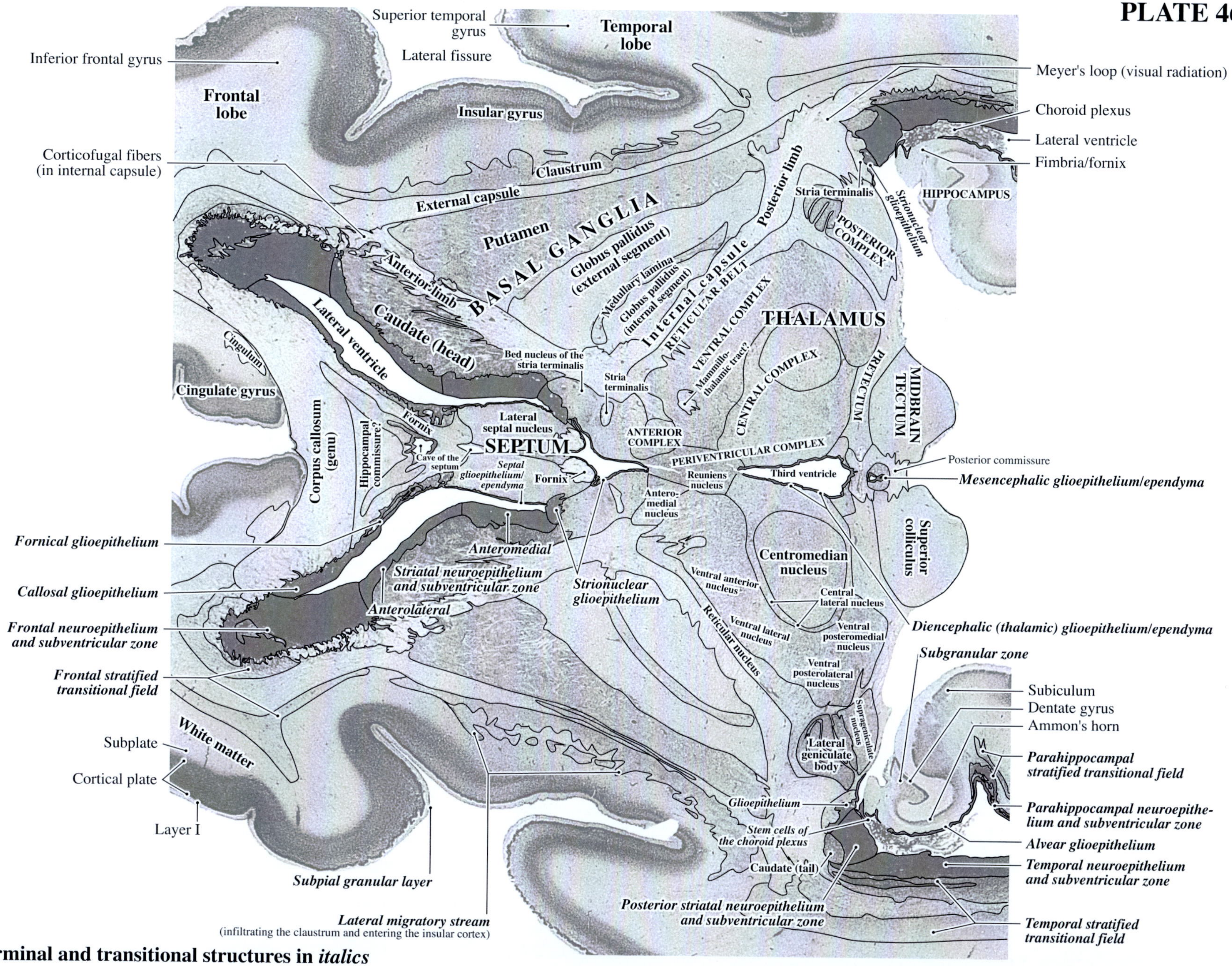

Germinal and transitional structures in *italics*

PLATE 47A
CR 235 mm
GW 26, Y16-59
Horizontal
Section 560

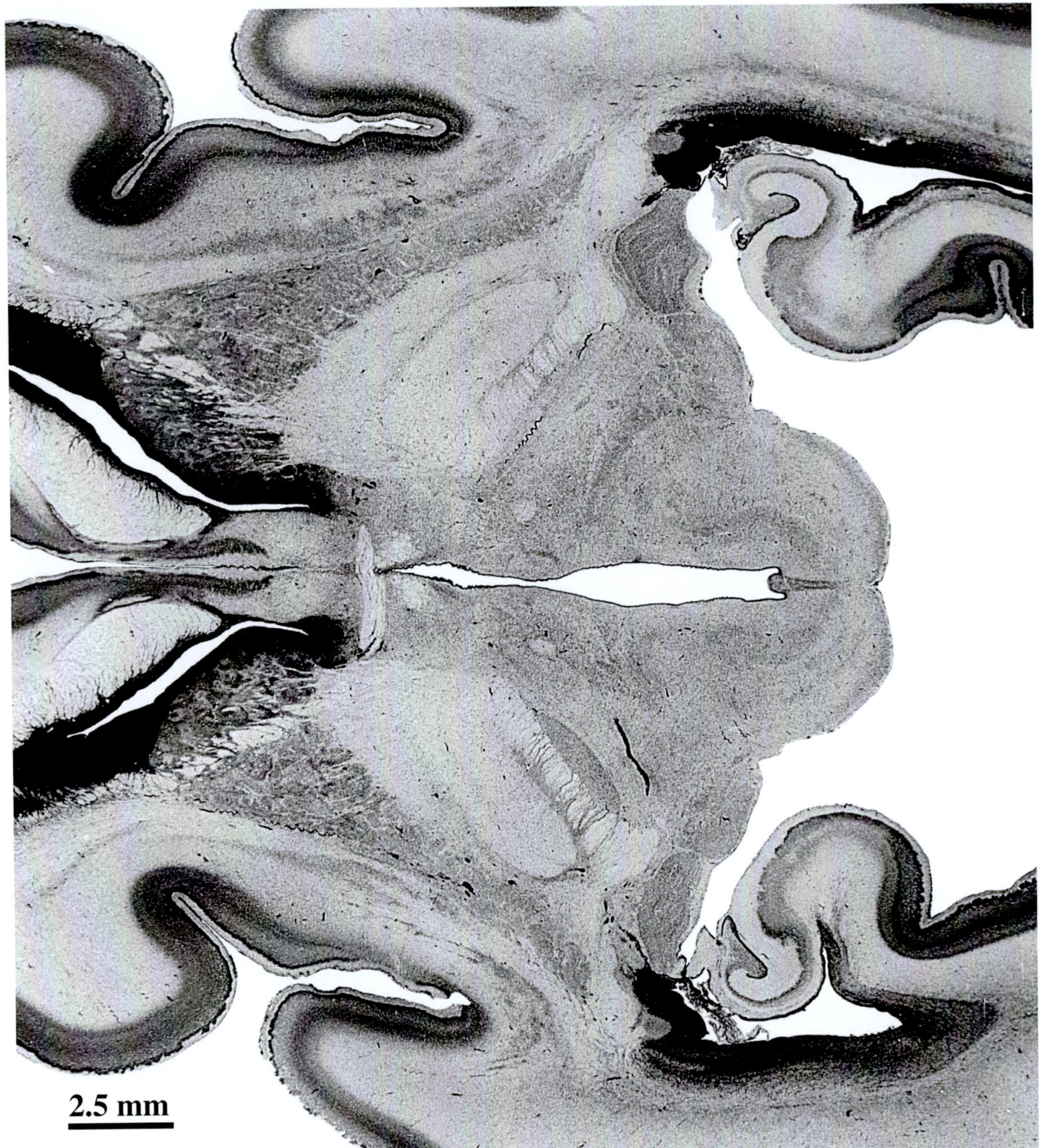

See the entire Section 560 in Plate 32.

PLATE 47B

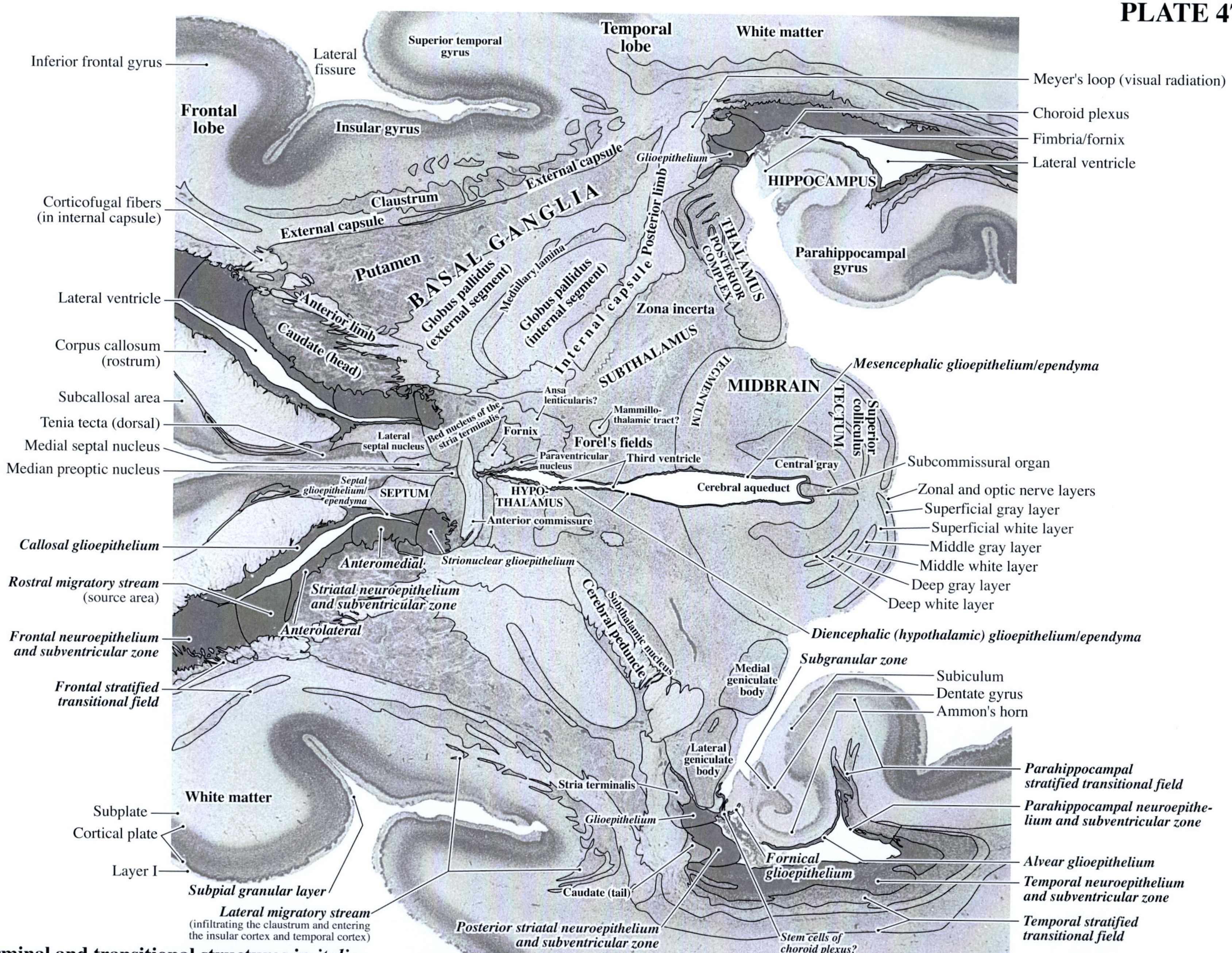

Germinal and transitional structures in *italics*

PLATE 48A, CR 235 mm, GW 26, Y16-59
Horizontal, Section 620

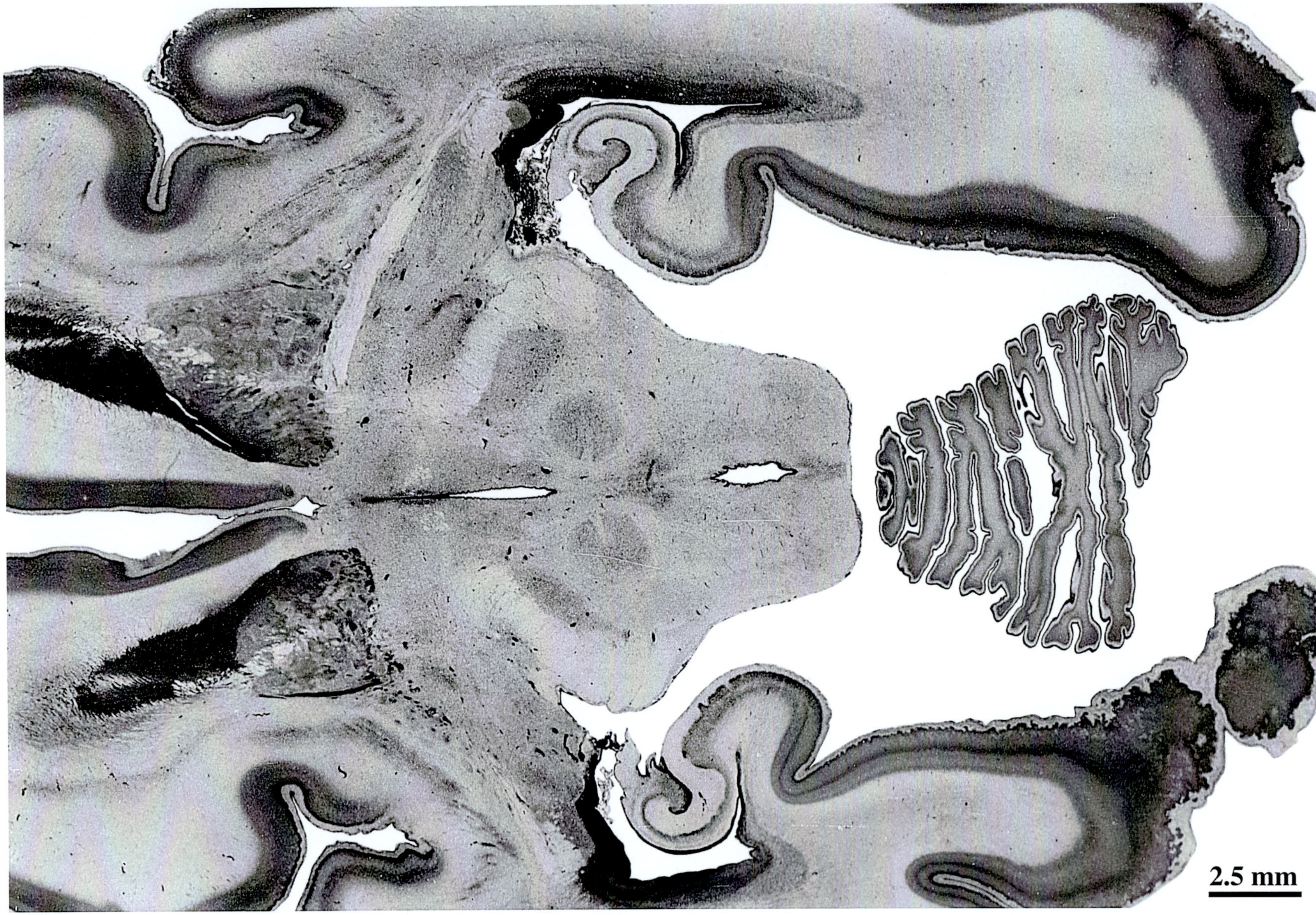

See the entire Section 620 in Plate 33.

PLATE 48B

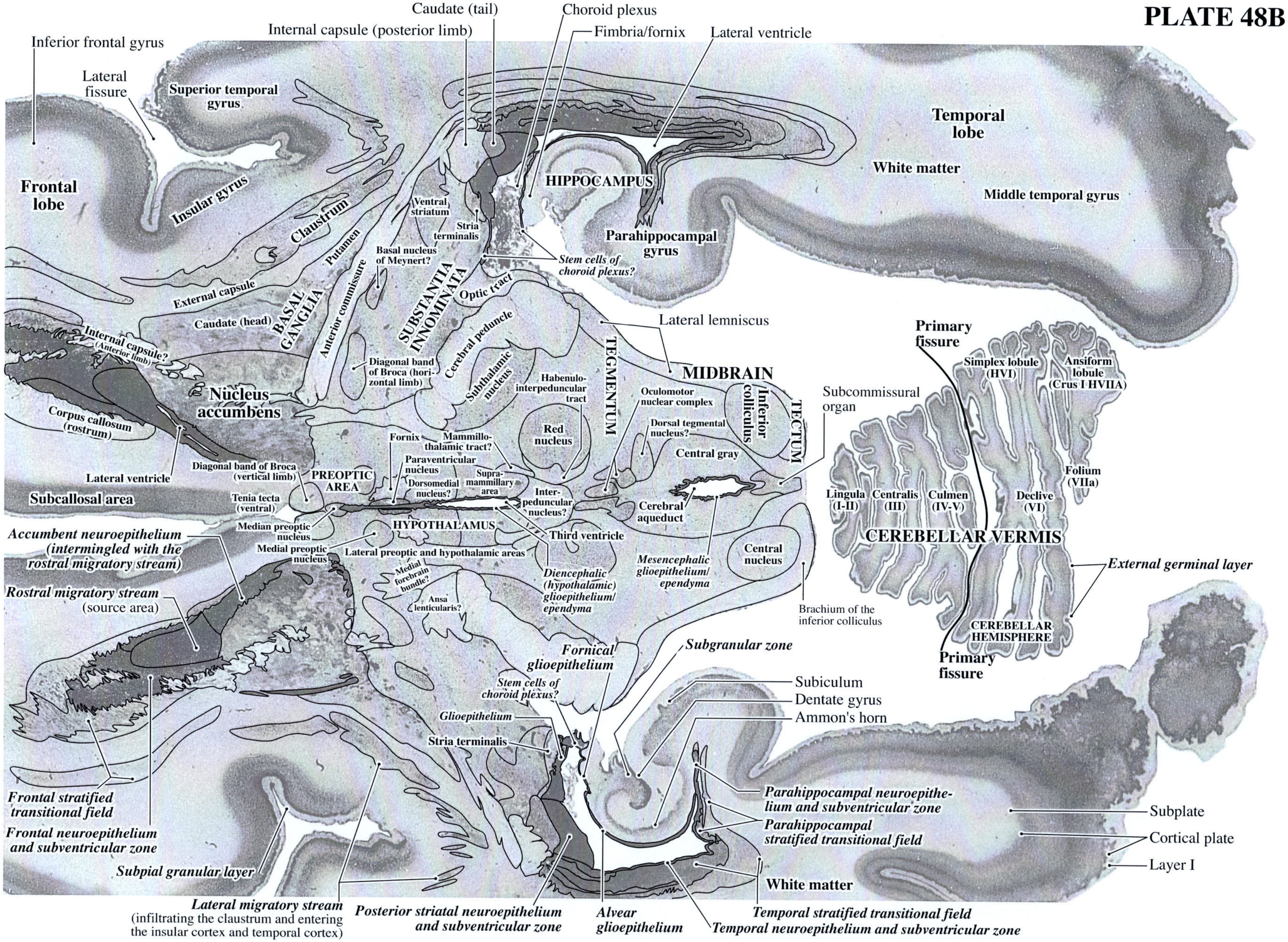

Germinal and transitional structures in *italics*

PLATE 49A, CR 235 mm, GW 26, Y16-59
Horizontal, Section 660

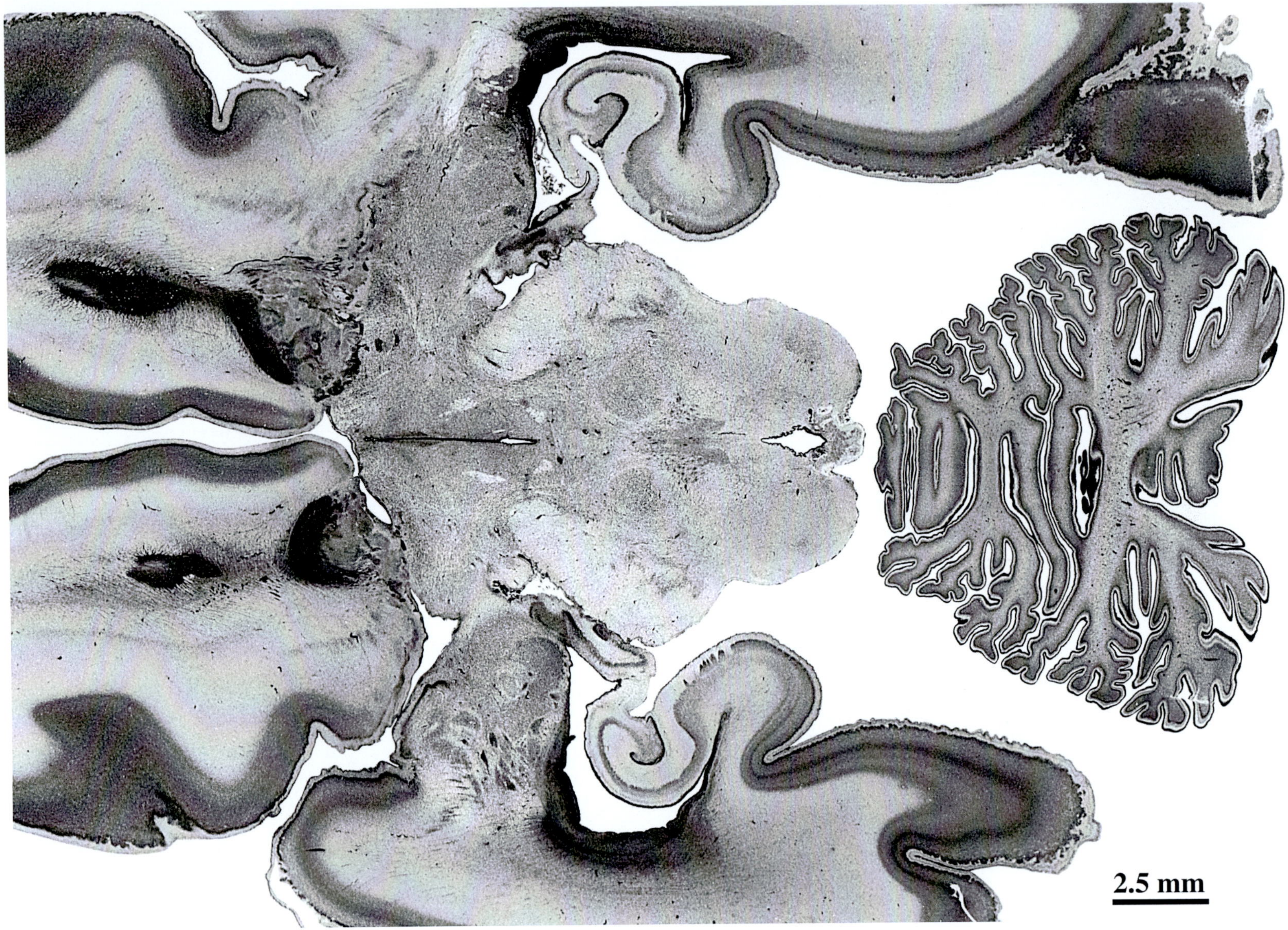

See the entire Section 660 in Plate 34.

PLATE 49B

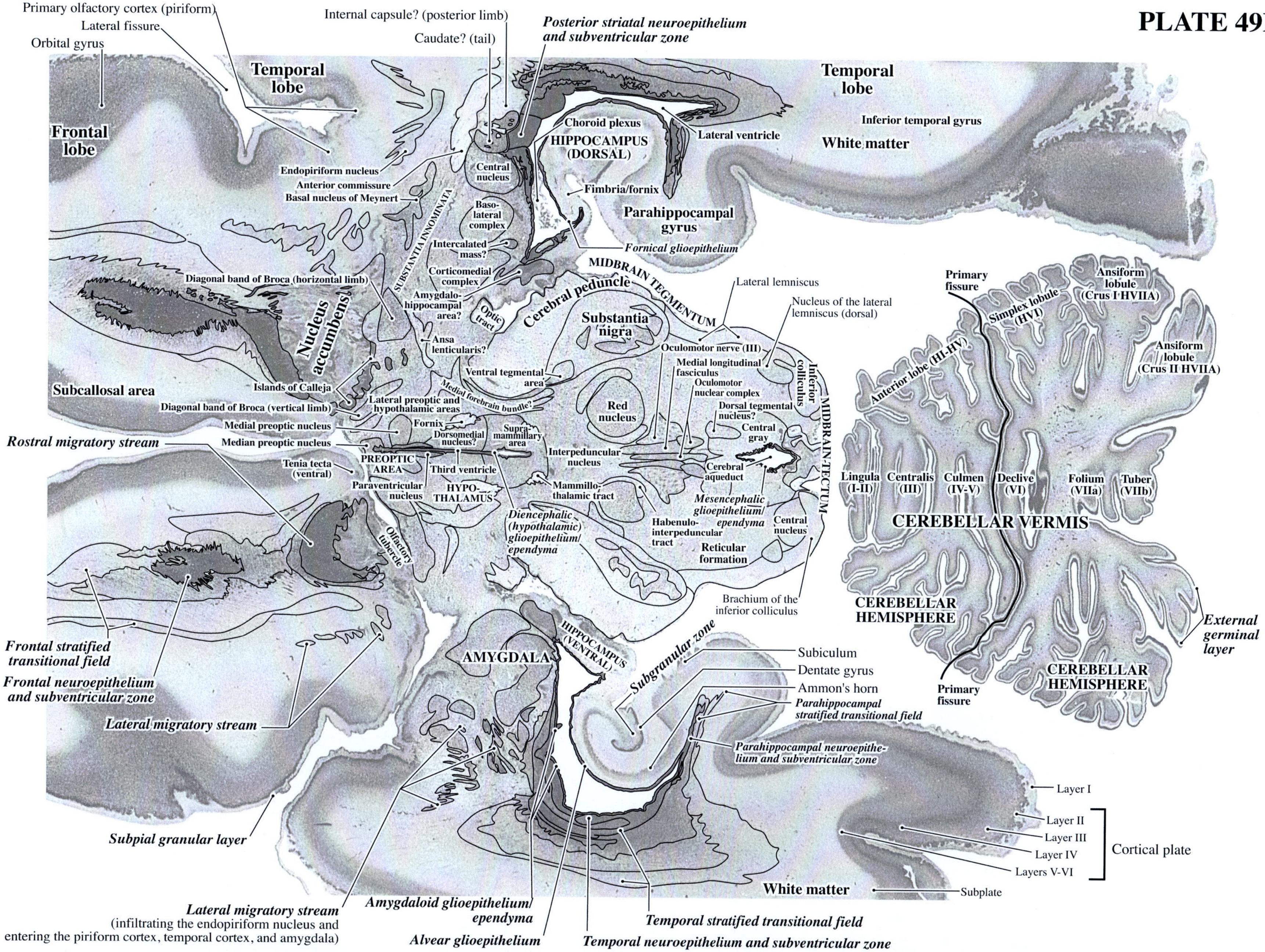

Germinal and transitional structures in *italics*

PLATE 50A, CR 235 mm, GW 26, Y16-59
Horizontal, Section 700

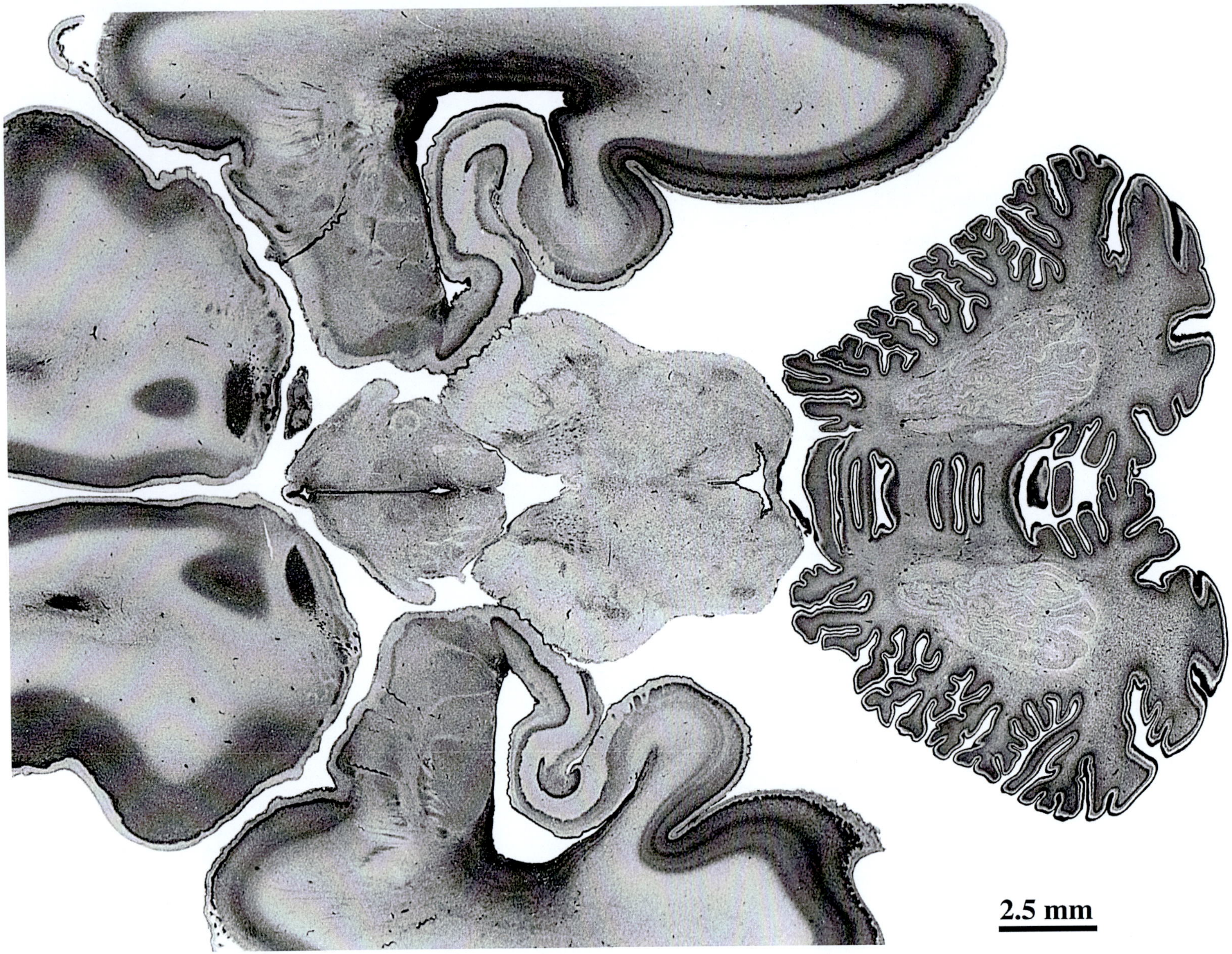

See the entire Section 700 in Plate 35.

PLATE 50B

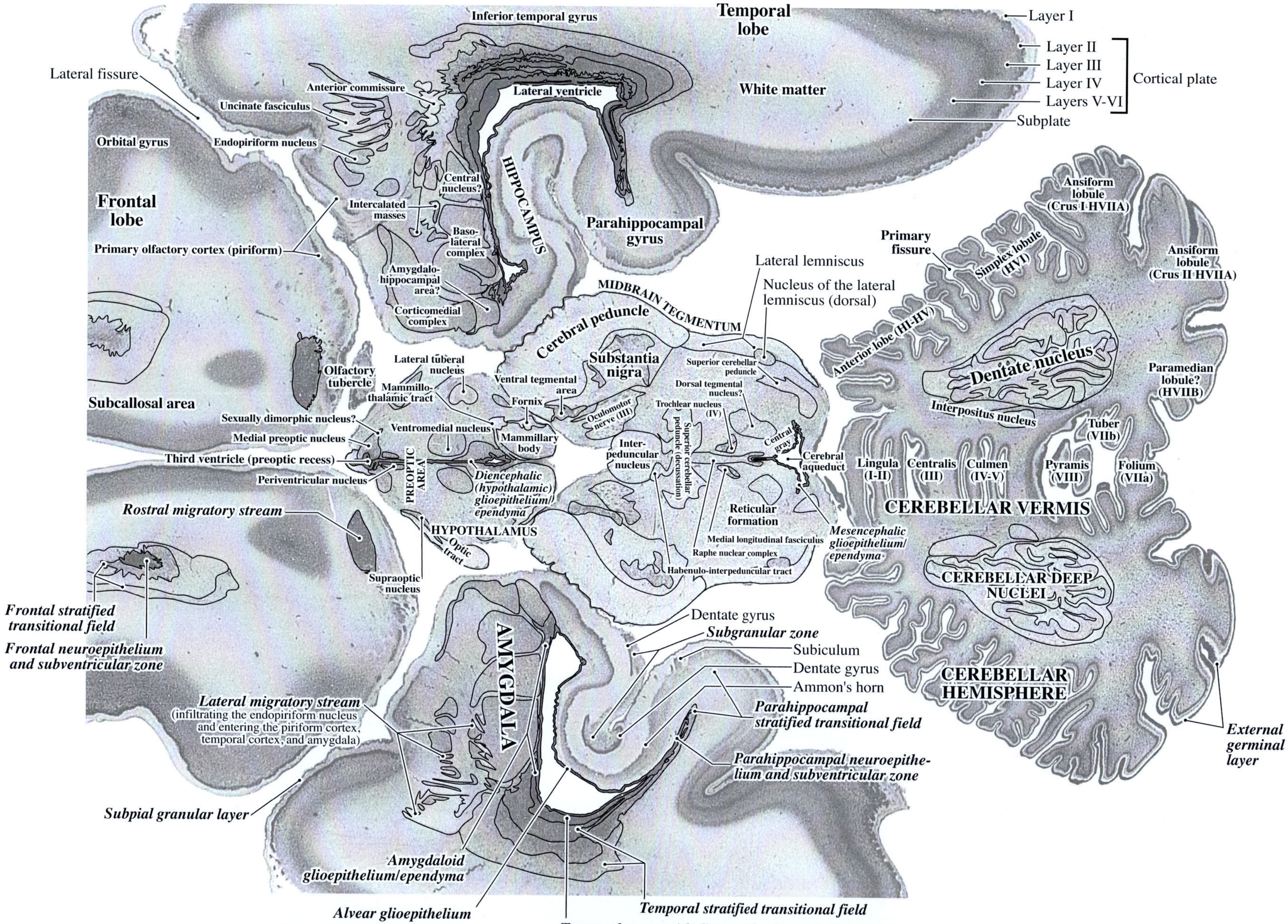

Germinal and transitional structures in *italics*

PLATE 51A, CR 235 mm, GW 26, Y16-59
Horizontal, Section 740

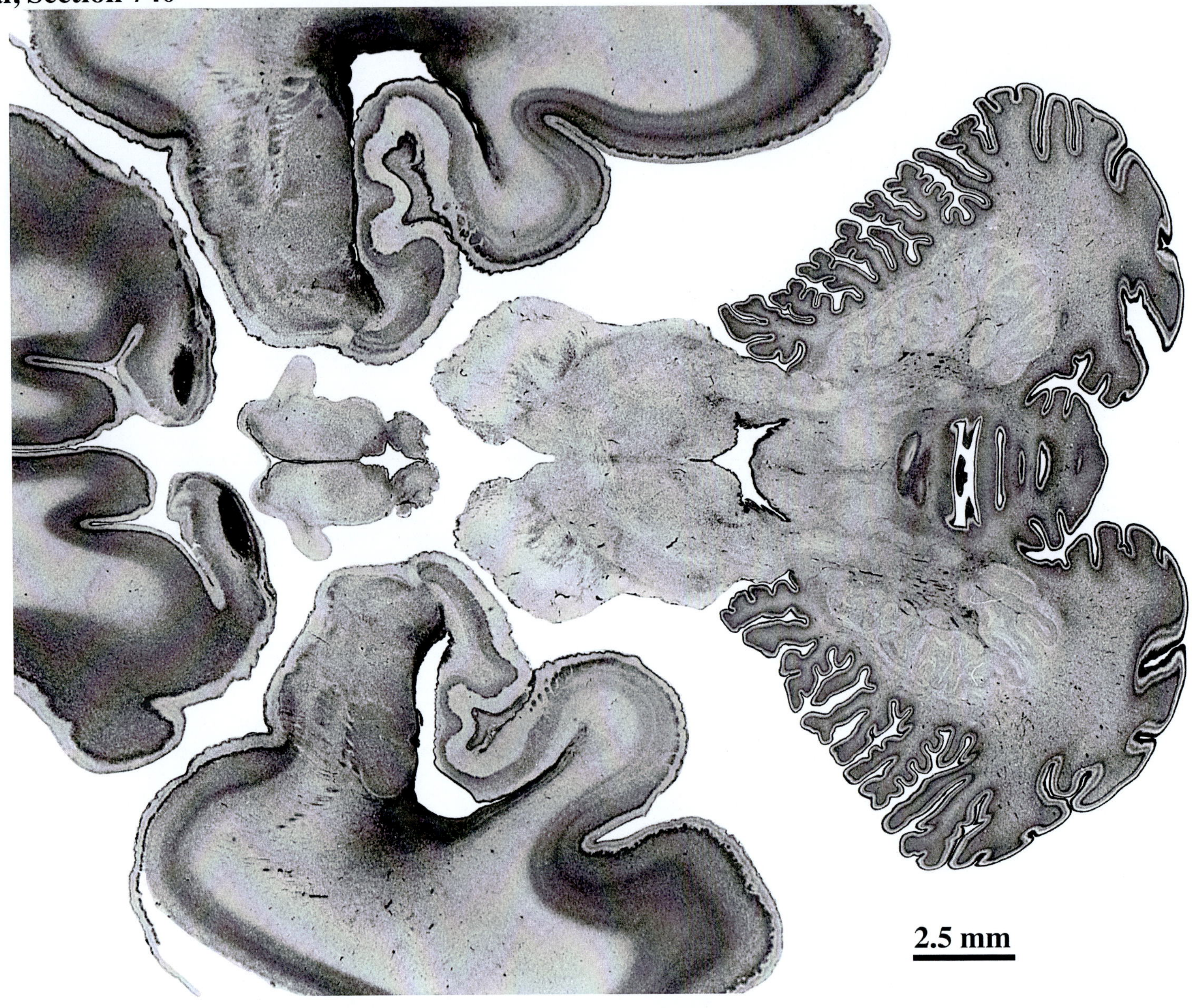

See the entire Section 740 in Plate 36.

PLATE 51B

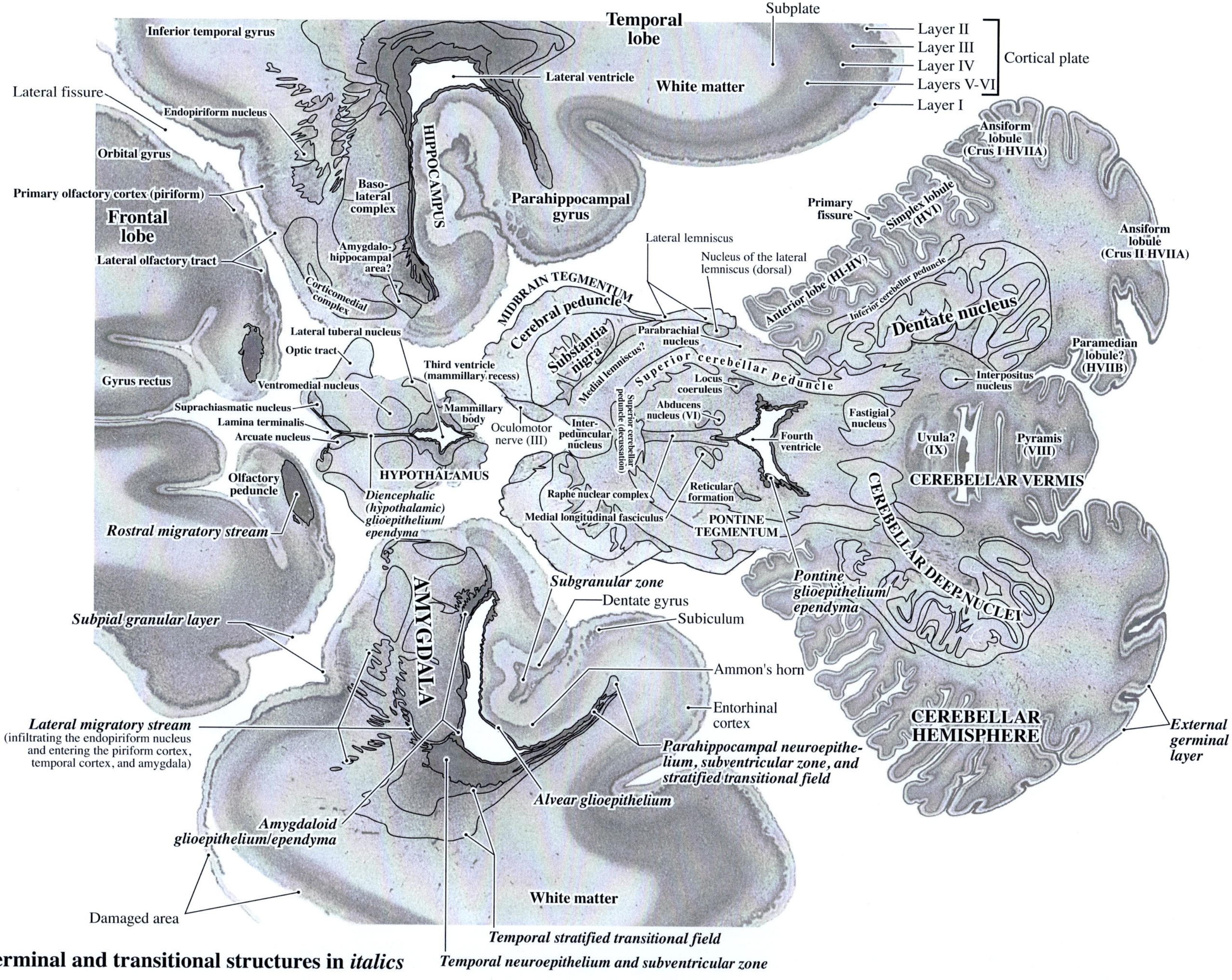

Germinal and transitional structures in *italics*

PLATE 52A, CR 235 mm, GW 26, Y16-59
Horizontal, Section 780

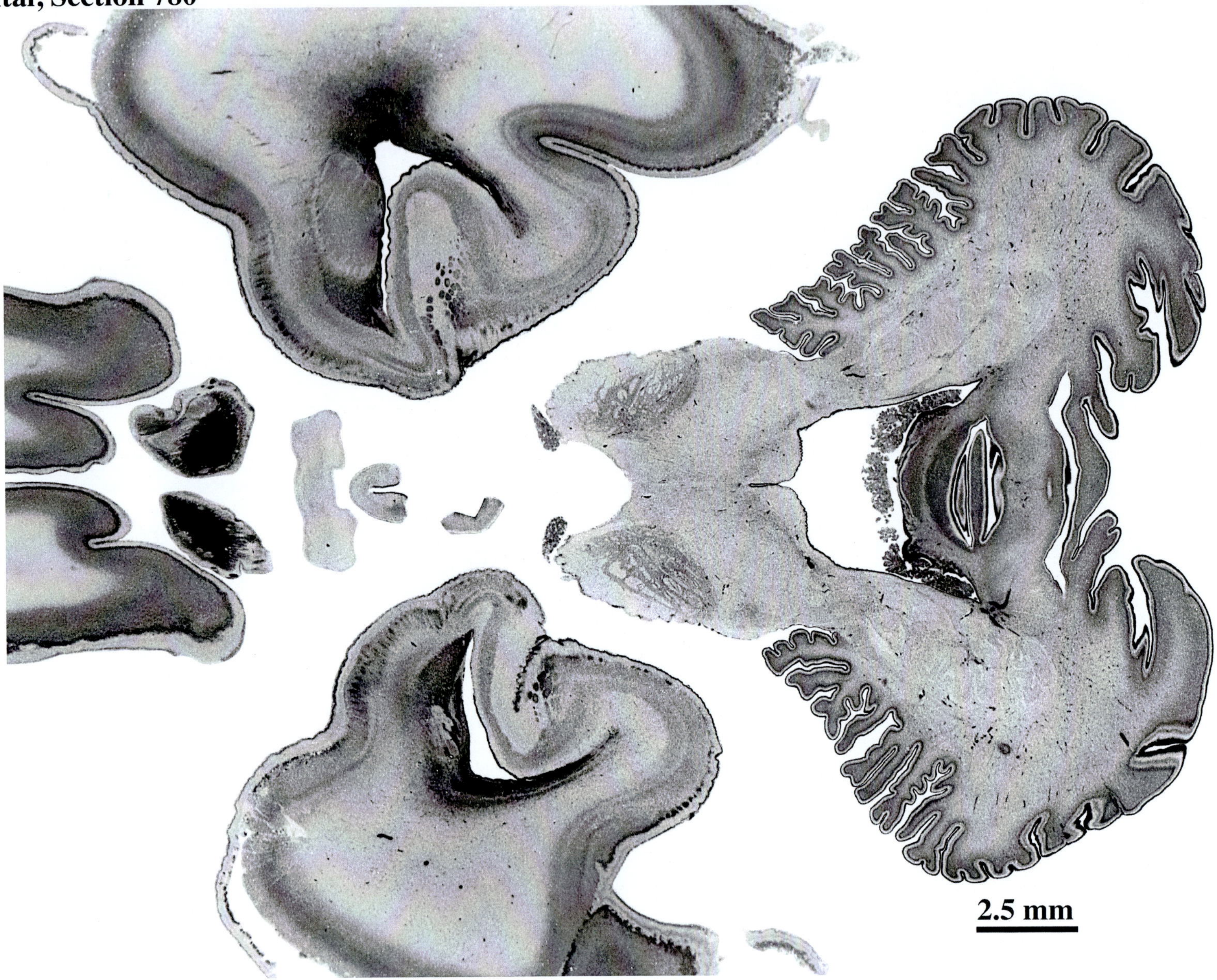

See the entire Section 780 in Plate 37.

PLATE 52B

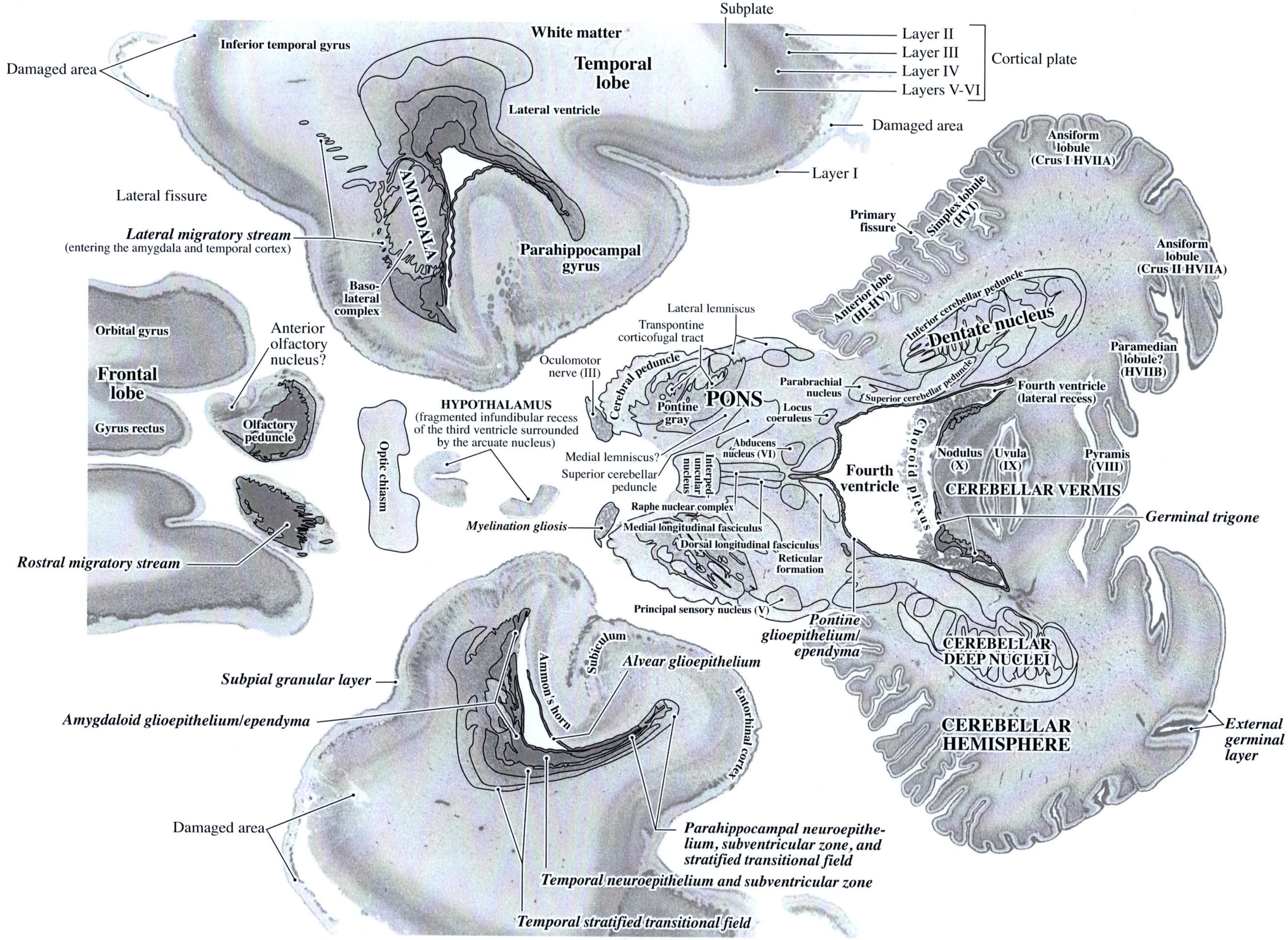

Germinal and transitional structures in *italics*

PLATE 53A, CR 235 mm, GW 26, Y16-59
Horizontal, Section 820

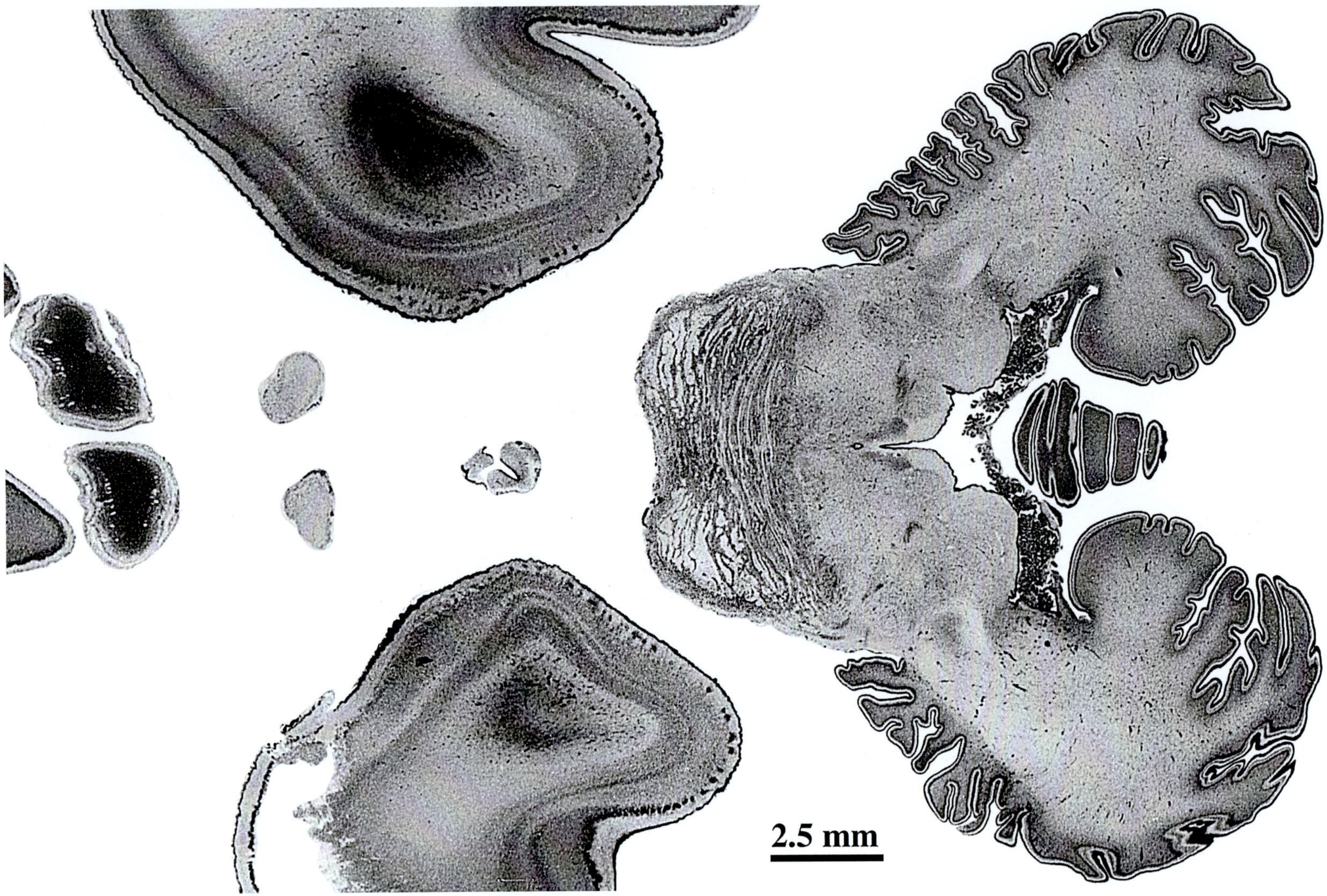

See the entire Section 820 in Plate 38.

PLATE 53B

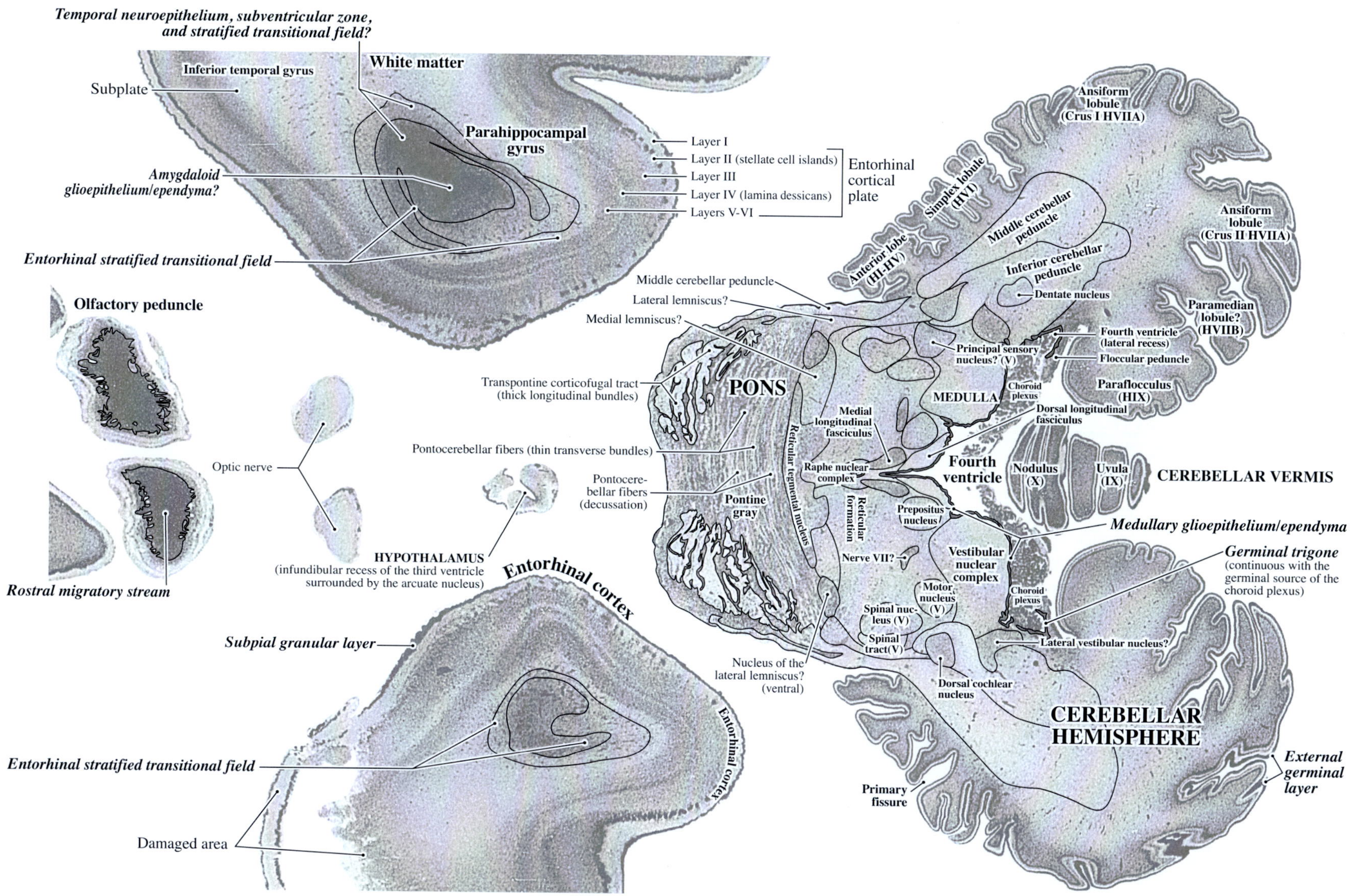

Germinal and transitional structures in *italics*

PLATE 54A
CR 235 mm
GW 26, Y16-59
Horizontal
Section 860

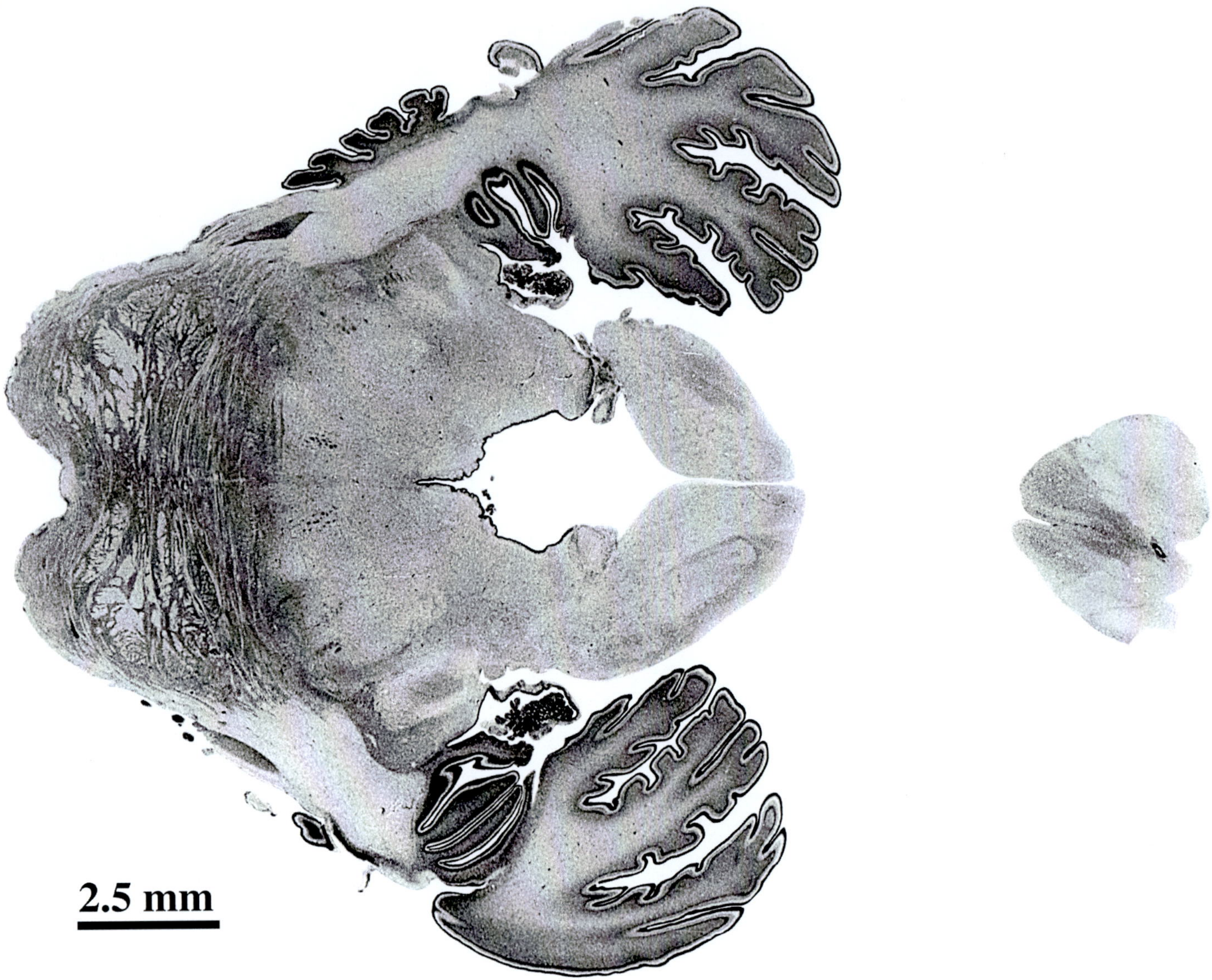

See the entire Section 860 in Plate 39.

PLATE 54B

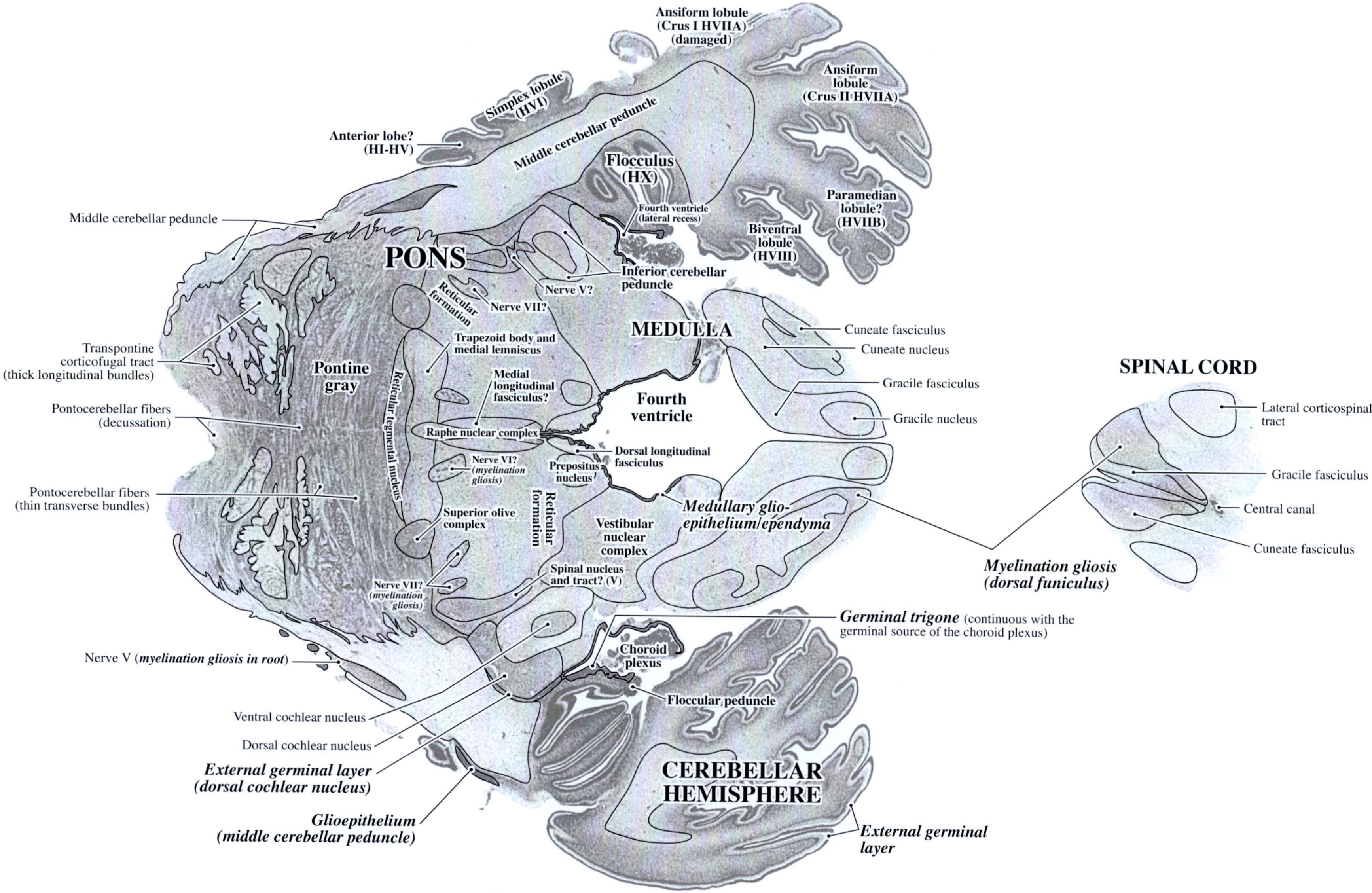

Germinal and transitional structures in *italics*

PLATE 55A
CR 235 mm
GW 26, Y16-59
Horizontal
Section 880

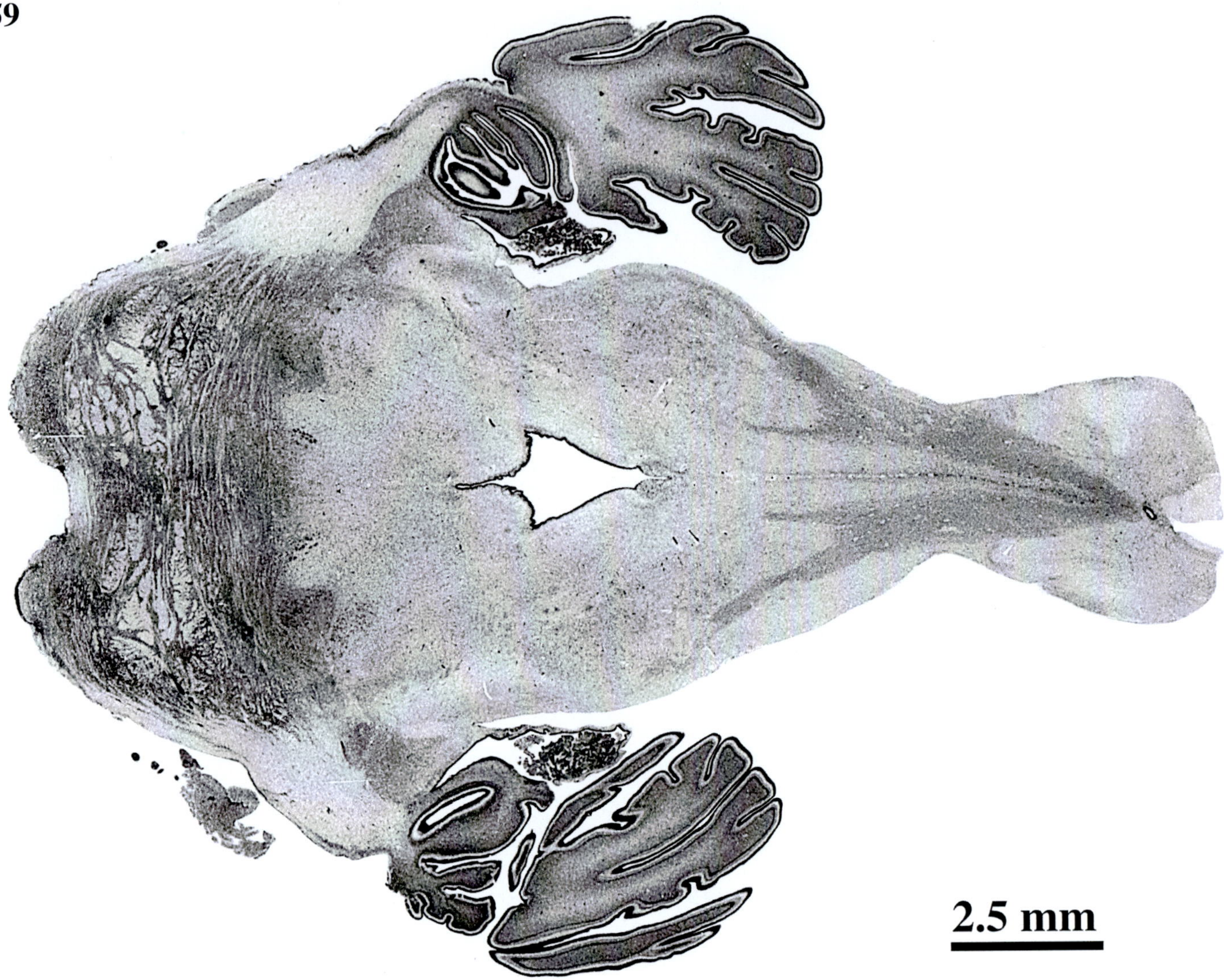

See the entire Section 880 in Plate 40.

PLATE 55B

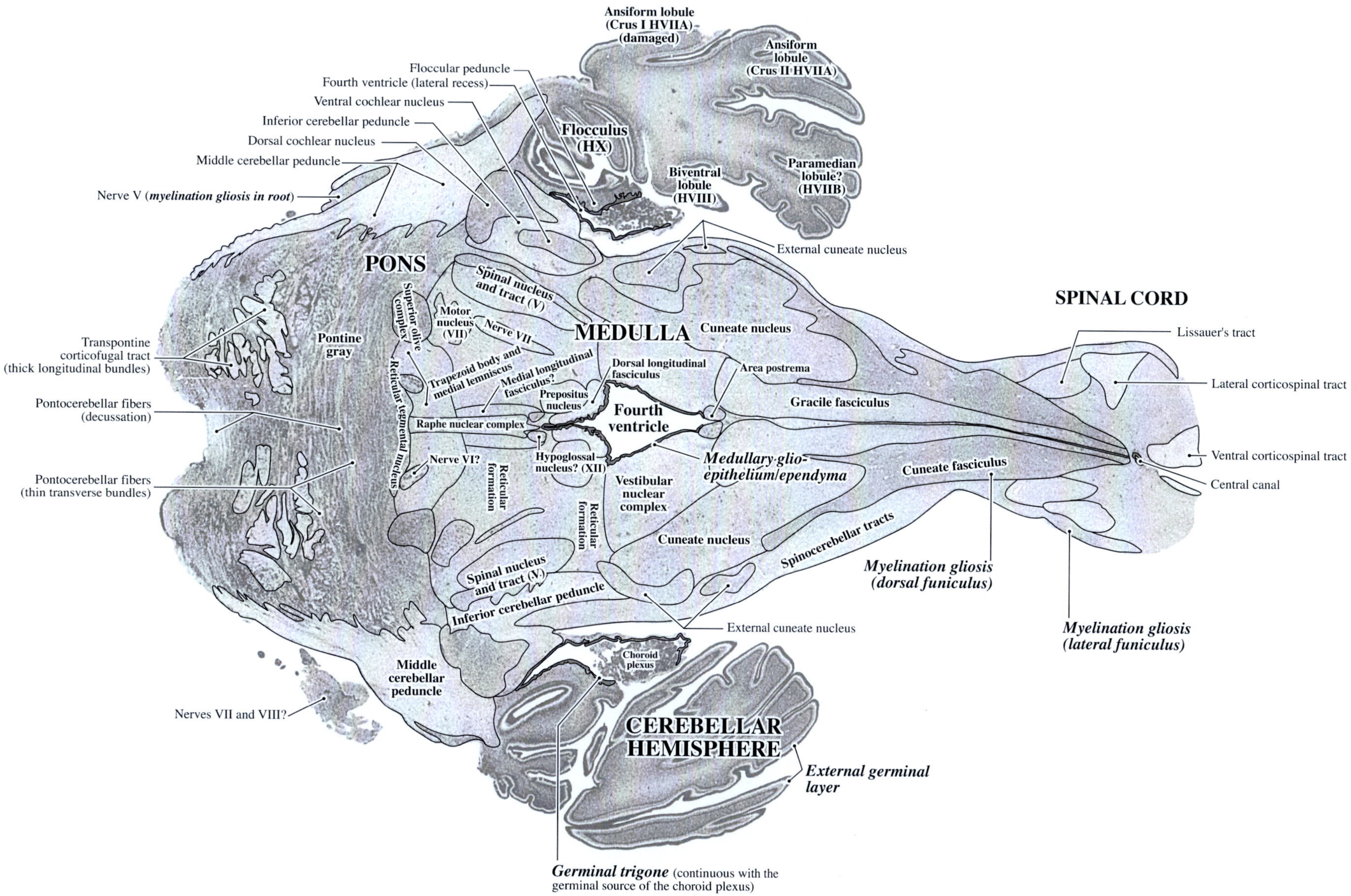

Germinal and transitional structures in *italics*

PLATE 56A
CR 235 mm
GW 26, Y16-59
Horizontal
Section 900

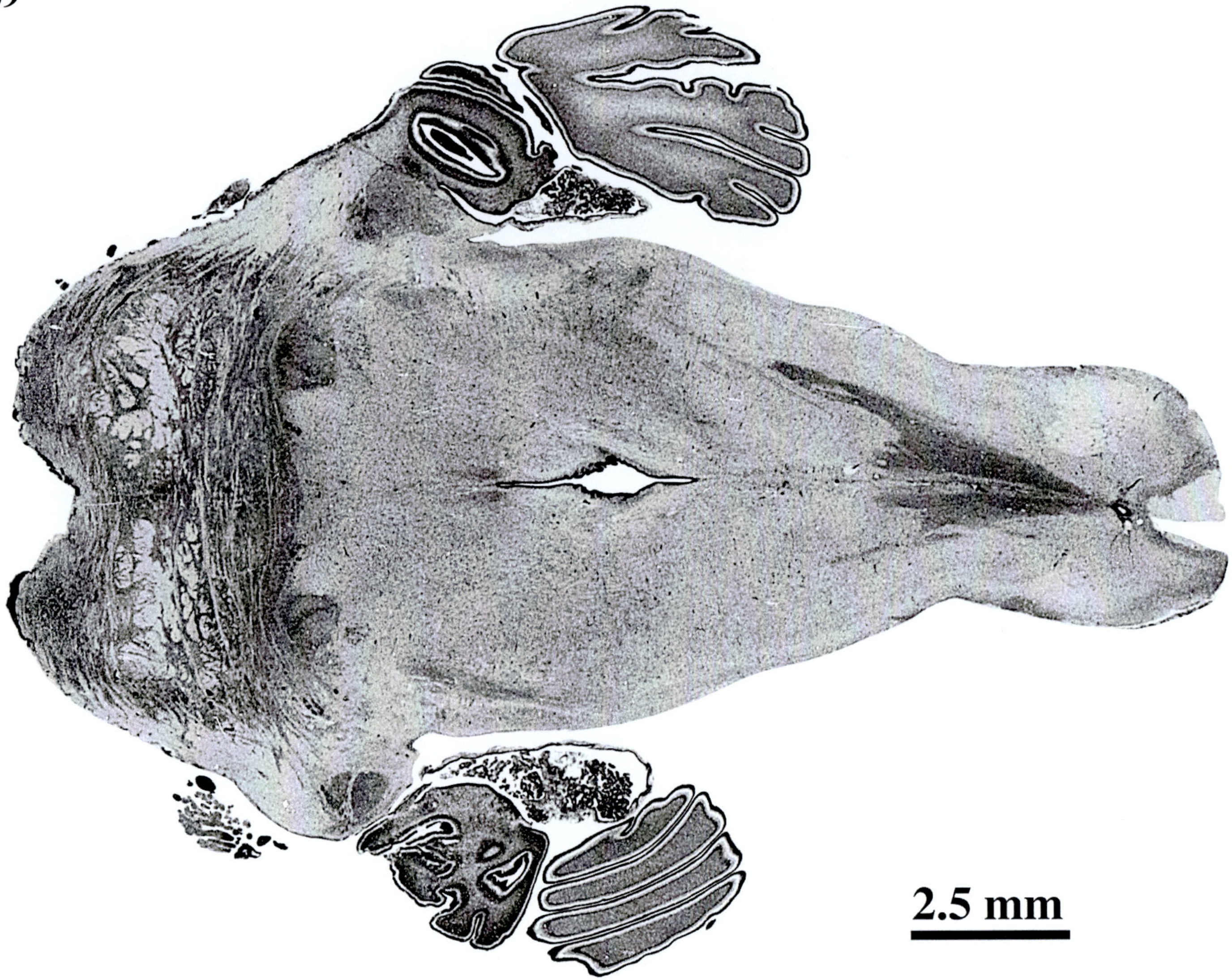

See the entire Section 900 in Plate 41.

PLATE 56B

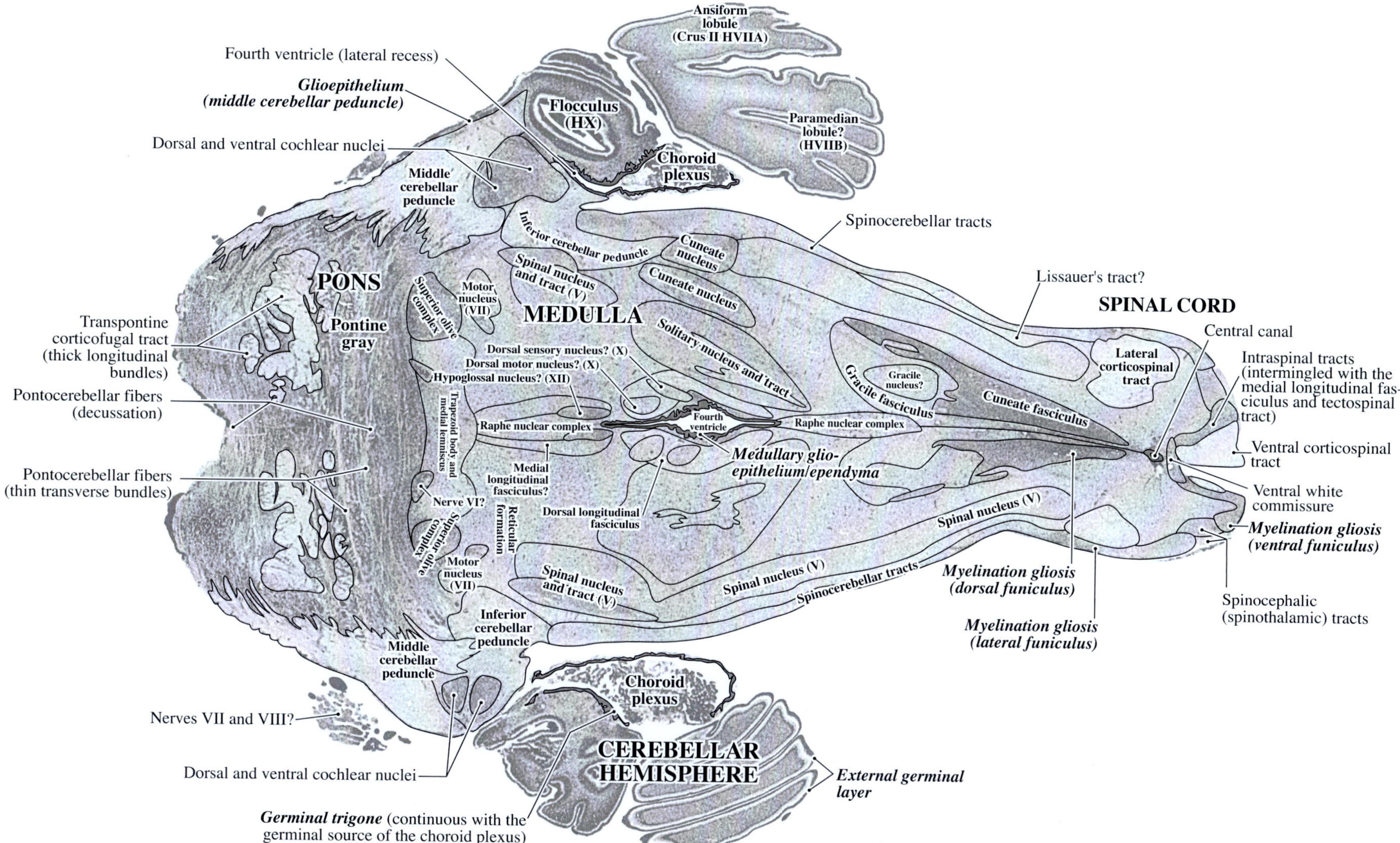

Germinal and transitional structures in *italics*

For Product Safety Concerns and Information please contact our EU
representative GPSR@taylorandfrancis.com Taylor & Francis Verlag GmbH,
Kaufingerstraße 24, 80331 München, Germany

Printed and bound by CPI Group (UK) Ltd, Croydon, CR0 4YY
14/09/2026
02216264-0001